Transient Flows: Mathematical Models, Laboratory Tests, Protection Elements and Systems

Transient Flows: Mathematical Models, Laboratory Tests, Protection Elements and Systems

Guest Editor

Vicente S. Fuertes-Miquel

MDPI

Basel • Beijing • Wuhan • Barcelona • Belgrade • Novi Sad • Cluj • Manchester

Guest Editor
Vicente S. Fuertes-Miquel
Departamento de Ingeniería
Hidráulica y Medio Ambiente
Universitat Politècnica de
València
Valencia
Spain

Editorial Office
MDPI AG
Grosspeteranlage 5
4052 Basel, Switzerland

This is a reprint of the Special Issue, published open access by the journal *Water* (ISSN 2073-4441), freely accessible at: https://www.mdpi.com/journal/water/special_issues/1KG42674K5.

For citation purposes, cite each article independently as indicated on the article page online and as indicated below:

Lastname, A.A.; Lastname, B.B. Article Title. *Journal Name* **Year**, *Volume Number*, Page Range.

ISBN 978-3-7258-4909-3 (Hbk)
ISBN 978-3-7258-4910-9 (PDF)
https://doi.org/10.3390/books978-3-7258-4910-9

Contents

About the Editor

Vicente S. Fuertes-Miquel

Vicente S. Fuertes-Miquel is an Industrial Engineer from the Universitat Politècnica de València, Spain (1992). He received his PhD in Hydraulic Engineering from the Department of Hydraulic and Environmental Engineering, Universitat Politècnica de València, Spain (2001). Currently, he is a Full Professor. His research is focused on the analysis, modeling, and design of hydraulic systems; the operation and maintenance of hydraulic systems; hydraulic elements (pumps, valves, pipelines, etc.); hydraulic transients in pressurized systems; hydraulic transients with entrapped air; the management of water services; the efficient use of water; and fluid facilities inside buildings. He is the author and editor of several books; he has written a number of research papers which have been published in high-impact journals; he has presented many of his contributions to international events; he has participated in numerous R&D projects and has served as the project coordinator on some of them; and he participates in doctoral programs and numerous postgraduate courses.

water

Editorial

Transient Flows: Mathematical Models, Laboratory Tests, Protection Elements and Systems

Vicente S. Fuertes-Miquel

Department of Hydraulic and Environmental Engineering, Universitat Politècnica de València, 46022 Valencia, Spain; vfuertes@upv.es

1. Introduction

Transient flows represent a critical topic in the field of hydraulic engineering, one that has witnessed significant progress in recent decades. These unsteady flow conditions arise from sudden changes in pipeline operations—whether due to intentional actions such as valve closures and pump starts or stops, or unexpected events such as power failures or system malfunctions. The consequences of poorly managed transients can be severe, leading to equipment failure, infrastructure damage, or even safety hazards [1]. As a result, the analysis and control of transient flows are now recognized as fundamental components in the design and operation of fluid transport systems.

Despite the considerable advancements made, transient flow analysis continues to pose scientific and engineering challenges. The transient behavior of fluid systems involves complex interactions between hydraulic phenomena, mechanical components, and operational variables [2]. To address this complexity, researchers have increasingly relied on developments in numerical modeling, computational fluid dynamics (CFD), and data-driven approaches.

This Special Issue aims to present recent advances in understanding, predicting, and mitigating transient flows. The selected contributions cover a broad range of topics, including mathematical modeling, laboratory experiments, CFD simulations, and real-world case studies. Particular emphasis is placed on the protective systems and devices used to prevent water hammer, the novel strategies employed to achieve transient control, hydraulic transients with air entrainment or water column separation, and assessments of the risk associated with these phenomena.

By compiling these diverse and high-quality contributions, this Special Issue provides readers with a comprehensive overview of current trends and future directions in transient flow research. We hope it serves as a valuable resource for researchers, engineers, and practitioners working to improve the resilience and efficiency of hydraulic systems under transient conditions.

2. Articles

In total, eight papers were published in this Special Issue, each contributing valuable insights into different aspects of transient flow analysis and control. The selected articles reflect the diversity and depth of current research in the field, ranging from advanced modeling techniques to practical applications in real-world systems. To provide a concise overview of the contents, Table 1 summarizes the titles, authors, and keywords of all contributions included in this Special Issue.

https://doi.org/10.3390/w17101454

Table 1. Summary of the papers published in the Special Issue "Transient Flows: Mathematical Models, Laboratory Tests, Protection Elements and Systems" for the journal Water.

Title	Authors	Keywords
Application of Newton–Raphson Method for Computing the Final Air–Water Interface Location in a Pipe Water Filling	· Dalia M. Bonilla-Correa · Óscar E. Coronado-Hernández · Vicente S. Fuertes-Miquel · Mohsen Besharat · Helena M. Ramos	· filling process · Newton–Raphson method · transient flow · pipelines · water
Smoothed Particle Hydrodynamics Simulations of Porous Medium Flow Using Ergun's Fixed-Bed Equation	· Carlos E. Alvarado-Rodríguez · Lamberto Díaz-Damacillo · Eric Plaza · Leonardo Di G. Sigalotti	· numerical methods · REV-scale simulation · pore-scale simulation · Ergun equation · packed beds · fluidized beds
Application of Machine Learning Coupled with Stochastic Numerical Analyses for Sizing Hybrid Surge Vessels on Low-Head Pumping Mains	· Ahmed M. A. Sattar · Abedalkareem Nedal Ghazal · Mohamed Elhakeem · Amgad S. Elansary · Bahram Gharabaghi	· water hammer · transient · machine learning · stochastic numerical analysis · flat profile
Analyzing Water Leakages in Parallel Pipe Systems with Rapid Regulating Valve Maneuvers	· Vicente S. Fuertes-Miquel · Alfonso Arrieta-Pastrana · Oscar E. Coronado-Hernández	· rigid water column model · mass oscillation equation · parallel pipes · valve maneuvers · water leakages · water distribution networks
Development of Pipeline Transient Mixed Flow Model with Smoothed Particle Hydrodynamics Based on Preissmann Slot Method	· Yixin Yang · Hexiang Yan · Shixun Li · Wenke Song · Fei Li · Huanfeng Duan · Kunlun Xin · Tao Tao	· transient mixed flow · Preissmann slot model · smoothed particle · pipeline systems
MOC-Z Model of Transient Cavitating Flow in Viscoelastic Pipe	· Giuseppe Pezzinga	· transients · cavitation · viscoelasticity · Kelvin–Voigt models · method of characteristics
Two-Dimensional Transient Flow in a Confined Aquifer with a Cut-Off Curtain Due to Dewatering	· Guangcheng Li · Huiming Lin · Min Deng · Lu Wang · Jianxiao Wang · Fanshui Kong · Yushan Zhang · Qinggao Feng	· 2D transient flow · semi-analytical solution · confined aquifer · cut-off curtain · sensitivity analysis

Table 1. *Cont.*

Title	Authors	Keywords
Influence of Entrapped Air on Hydraulic Transients During Rapid Closure of a Valve Located Upstream and Downstream of an Air Pocket in Pressurised Pipes	Oscar Pozos-Estrada	· hydraulic transients · rapid closure · hydraulic jump · air pocket · air bubbles

The study of thermodynamic behavior during filling processes in water pipes with trapped air is complex. Using the rigid water column (RWC) model, Bonilla-Correa et al. (Contribution 1) propose a computationally efficient method that is able to estimate critical variables such as the air pocket pressure and water column length. A Newton–Raphson-based formulation enables the direct computation of these values. The proposed method is validated against experimental data and shows strong agreement with a detailed mathematical model. A sensitivity analysis identifies the most influential parameters affecting the final system state.

Alvarado-Rodríguez et al. (Contribution 2) investigate fluid flow through a porous layer in a rectangular channel using Smoothed Particle Hydrodynamics (SPH) at both the pore scale and the Representative Elementary Volume (REV) scale, applying the Ergun equation. The study compares velocity profiles and pressure losses for different grain configurations. The results show that REV and pore-scale predictions align best at intermediate porosities (0.44–0.55). Significant discrepancies arise at higher porosities or non-uniform arrangements, highlighting the influence of grain geometry on flow behavior and the accuracy of the model.

Sattar et al. (Contribution 3) address surge protection challenges in low-head pipelines during pump failures, proposing a novel hybrid surge vessel with a dipping tube as an efficient solution. A stochastic numerical approach combined with machine learning—specifically genetic programming—is used to develop a predictive model for optimal vessel sizing. The model, validated with 2000 simulated cases, accurately estimates air and compression chamber volumes and offers a classification index for economical applications. The results outperform traditional design charts and highlight key influencing parameters such as the pipe diameter and static head.

Water leakages in water distribution networks are a topic of great interest. Fuertes-Miquel et al. (Contribution 4) introduce a novel mathematical model based on the mass oscillation equation to improve the prediction of water leakages in distribution systems, particularly under the influence of regulating valves. Unlike traditional Bernoulli-based approaches, this model captures the dynamic effects of system inertia. The method is applicable to parallel pipe systems and is validated through a case study involving two parallel pipelines. The results highlight increased leakage volumes in parallel configurations compared to single-pipe systems, offering engineers a more accurate tool for leak assessment and infrastructure planning.

Yang et al. (Contribution 5) introduce a novel one-dimensional numerical model that combines the Smoothed Particle Hydrodynamics (SPH) method with the Preissmann Slot Method (PSM) to simulate transient mixed flows in pipeline systems. The SPH-PSM model offers enhanced capabilities for capturing complex free-surface and pressurized flow interactions. After optimizing the empirical parameters, the model is validated against various flow regimes and shows strong agreement with the reference data. The results confirm the model's effectiveness as a robust alternative for analyzing and understanding mixed-flow phenomena in urban drainage and pipeline systems.

Pezzinga (Contribution 6) presents a unitary method for solving transient cavitating flow in viscoelastic pipes, using the Method of Characteristics (MOC) combined with a Z-mirror numerical scheme (MOC-Z model). The approach accommodates both viscoelasticity and cavitation, extending the MOC-Z for Courant numbers less than 1 to resolve numerical instabilities. Four viscoelastic models (KV, GKV1, GKV2 and GKV3) are tested, and the results show that the KV model outperforms the more complex GKV models in several cases. This study provides insights into the optimal model selection for different flow conditions.

Another topic is the transient flow of groundwater in an aquifer. Li et al. (Contribution 7) present a two-dimensional analytical model of transient groundwater flow in a confined aquifer with a cut-off curtain, which is commonly used in foundation pit dewatering. The model considers both the dewatering well and the cut-off curtain in an anisotropic aquifer. Using Laplace and Fourier cosine transformations, the authors derive a semi-analytical solution for drawdown, which is validated against numerical data. A sensitivity analysis reveals that drawdown is highly influenced by the pumping rate, aquifer properties, and well structure, with specific sensitivities inside and outside the curtain.

Hydraulic transients with entrapped air are a particular case in which there are two fluids (water and air) in two phases (liquid and gas). Pozos-Estrada (Contribution 8) explores the dual impact of trapped air on fluid transients in pressurized pipelines, particularly during rapid valve closure. The study examines how air pockets and dispersed bubbles affect transient pressures, using both experiments and numerical simulations. The results show that air can mitigate pressure transients when the valve is downstream but exacerbate them when positioned upstream. This highlights the importance of considering entrapped air during pipeline design to optimize the management of transient pressure and prevent adverse effects, depending on the valve and pipeline configuration.

3. Conclusions

This Special Issue comprises a diverse set of studies that focus on various aspects of hydraulic transients, particularly those influenced by the presence of entrapped air in pipeline systems. The collection of papers emphasizes the significant role that transient phenomena, such as pressure surges, cavitation, and water hammer, play in the design and operation of fluid transport systems. Several of the articles specifically address the complexities of simulating two-phase flows, with a focus on understanding and mitigating the effects of air pockets and gas entrainment during transient events. These studies offer novel numerical models and experimental data that contribute to a more comprehensive understanding of transient dynamics, providing valuable insights into mitigating the risks associated with air-induced pressure variations.

One of the key contributions of this Special Issue is the exploration of computational fluid dynamics (CFD) models that are designed to simulate hydraulic transients during both filling and emptying processes in water pipelines. Some of the articles delve into the development and application of CFD models, offering a deeper understanding of flow behavior and the influence of pipeline geometry on transient responses. These models, validated through experimental testing, serve as powerful tools for engineers seeking to optimize the design and performance of pipeline systems under dynamic conditions.

Another important topic discussed is the characterization and modeling of air valves, which play a crucial role in managing transient pressures in pipeline systems. The research presented in this Special Issue highlights the need for accurate air valve sizing and effective strategies for controlling air pockets to prevent negative pressure surges and other potentially hazardous phenomena. The studies underscore the necessity of incorporating air valve dynamics into the overall design process to ensure the stability and safety of hydraulic systems.

Despite the advances in understanding hydraulic transients, it is clear that many aspects of this field require further investigation. The interaction between transient flow events and entrapped air, the development of more accurate numerical models, and the optimization of air management strategies remain ongoing challenges. As pipeline systems grow in complexity and scale, the need for more advanced, reliable methods for simulating and controlling transient phenomena becomes increasingly critical. Future research should continue to focus on the refinement of existing models, the enhancement of computational techniques, and the exploration of new solutions to improve the safety, efficiency, and resilience of fluid transport systems.

Funding: This research received no external funding.

Acknowledgments: Thanks to all of the contributions to this Special Issue, the time invested by each author, as well as the anonymous reviewers who contributed to the development of the articles in this Special Issue. The Guest Editor is very pleased with the review process and management of the Special Issue.

Conflicts of Interest: The author declares no conflicts of interest.

List of Contributions:

1. Bonilla-Correa, D.M.; Coronado-Hernández, O.E.; Fuertes-Miquel, V.S.; Besharat, M.; Ramos, H.M. Application of Newton–Raphson Method for Computing the Final Air–Water Interface Location in a Pipe Water Filling. *Water* **2023**, *15*, 1304.
2. Alvarado-Rodríguez, C.E.; Díaz-Damacillo, L.; Plaza, E.; Sigalotti, L.G. Smoothed Particle Hydrodynamics Simulations of Porous Medium Flow Using Ergun's Fixed-Bed Equation. *Water* **2023**, *15*, 2358.
3. Sattar, A.M.A.; Ghazal, A.N.; Elhakeem, M.; Elansary, A.S.; Gharabaghi, B. Application of Machine Learning Coupled with Stochastic Numerical Analyses for Sizing Hybrid Surge Vessels on Low-Head Pumping Mains. *Water* **2023**, *15*, 3525.
4. Fuertes-Miquel, V.S.; Arrieta-Pastrana, A.; Coronado-Hernández, O.E. Analyzing Water Leakages in Parallel Pipe Systems with Rapid Regulating Valve Maneuvers. *Water* **2024**, *16*, 926.
5. Yang, Y.; Yan, H.; Li, S.; Song, W.; Li, F.; Duan, H.; Xin, K.; Tao, T. Development of Pipeline Transient Mixed Flow Model with Smoothed Particle Hydrodynamics Based on Preissmann Slot Method. *Water* **2024**, *16*, 1108.
6. Pezzinga, G. MOC-Z Model of Transient Cavitating Flow in Viscoelastic Pipe. *Water* **2024**, *16*, 1610.
7. Li, G.; Lin, H.; Deng, M.; Wang, L.; Wang, J.; Kong, F.; Zhang, Y.; Feng, Q. Two-Dimensional Transient Flow in a Confined Aquifer with a Cut-Off Curtain Due to Dewatering. *Water* **2025**, *17*, 601.
8. Pozos-Estrada, O. Influence of Entrapped Air on Hydraulic Transients During Rapid Closure of a Valve Located Upstream and Downstream of an Air Pocket in Pressurised Pipes. *Water* **2025**, *17*, 927.

References

1. Chaudhry, M.H. *Applied Hydraulic Transients*, 3rd ed.; Springer: New York, NY, USA, 2014.
2. Thorley, A.R.D. *Fluid Transients in Pipeline Systems*, 2nd ed.; Professional Engineering Publishing: London, UK, 2004.

water

MDPI

Article

Analyzing Water Leakages in Parallel Pipe Systems with Rapid Regulating Valve Maneuvers

Vicente S. Fuertes-Miquel [1], Alfonso Arrieta-Pastrana [2] and Oscar E. Coronado-Hernández [2,*]

1 Departamento de Ingeniería Hidráulica y Medio Ambiente, Universitat Politècnica de València, 46022 Valencia, Spain; vfuertes@upv.es

2 Instituto de Hidráulica y Saneamiento Ambiental, Universidad de Cartagena, Cartagena de Indias 130001, Colombia

* Correspondence: ocoronadoh@unicartagena.edu.co

Abstract: Water utilities face the challenge of addressing physical leaks generated from the aging of water distribution systems and the need for more innovative practices to manage water infrastructure efficiently. Water leakages are typically modeled using extended period simulations based on Bernoulli's equation. However, this approach must be revised since traditional methods do not appropriately simulate variations induced by regulating valves. In this study, the authors developed a mathematical model based on the mass oscillation equation, which is well-suited for predicting water leakages while accounting for system inertia from regulating valves. This approach is versatile and can be applied to all parallel pipe systems. A comprehensive practical application involving two parallel pipes has been conducted. The aim is to provide engineers and designers with a tool to assess the total volume of water leaks caused by regulating valves in real-world water distribution networks. Furthermore, the study includes a comparative analysis with a single pipe configuration to illustrate how parallel systems lead to increased leaks in contrast to simpler pipe setups.

Keywords: rigid water column model; mass oscillation equation; parallel pipes; valve maneuvers; water leakages; water distribution networks

1. Introduction

Water leakages are a current issue faced by developed and developing countries worldwide. They account for a significant portion of the total water loss in water distribution systems and constitute physical water loss. Water leakages can be identified in pipelines, household connections, and seepage from water storage tanks [1,2]. Furthermore, leaks occur at the installation sites of control, measurement, and safety equipment [2]. Leakages tend to be more pronounced in developing countries when compared to developed ones. According to data from IBNET [3], water leakages in certain South American countries can reach as high as 30% or more, whereas in Australia and the United States of America, they exceed 10%. The higher the incidence of water leakages, the greater the expenditure water utilities incur for water production, resulting in higher user costs. In this context, two Sustainable Development Goals (SDGs) are interconnected with water leakages [4,5]: (i) ensuring access to clean water and sanitation and (ii) promoting sustainable cities and communities. Water leakages result from various factors, including the aging infrastructure leading to material deterioration over time, the utilization of deficient technologies, inadequate pressure management programs, improper maintenance of water infrastructure, and adherence to local regulatory frameworks [6–8].

At present, water utilities are addressing the issue of water leakages through the implementation of digital twin technology, which comprises several layers [9–12]:

- A Geographical Information System (GIS) that describes water infrastructure;
- Sensors to measure hydraulic variables in the system;

- Supervisory Control and Data Acquisition (SCADA) systems collect data from sensors placed in situ and transmit it to a central station through a communication system;
- Smart Metering for network control and customer service;
- A computerized maintenance management system;
- A hydraulic model designed to simulate water leakages in water distribution systems.

Additionally, artificial intelligence techniques to predict future water losses in water installations have been implemented in water distribution systems [13,14]. The greater the water leakages, the higher the energy expended in production [15,16].

Hydraulic models are crucial in assessing and mitigating water leakages, forming the foundation for robust digital twin models. In 1990, Jowitt and Xu [17] introduced an algorithm based on the Hazen–Williams formula and linear theory methods to quantify and minimize water leakages in water distribution systems. Similarly, Giustolisi, Savic, and Kapelan [18] developed a steady-state model for simulating water distribution systems, utilizing classical hydraulic equations, pressure-driven demand, and accounting for leakages, including elevation considerations. Water leakages are evaluated using steady-state formulations, such as EPANET 2.2 or other commercial software packages, in conjunction with a water balance framework [19–21].

While the steady-state approach, also known as extended period simulation, is commonly used by water utilities to calculate water leak volumes, it falls short when system inertia must be factored into the calculations. This is particularly relevant when regulating valves operating within water distribution systems. When a regulating valve is manipulated, it introduces specific oscillations not captured by extended period simulations, potentially leading to underestimation or overestimation of water leak volumes. Consequently, Coronado-Hernández et al. (2023) [22] introduced a mathematical model to compute water leakages in single pipelines, considering system inertia.

Inertial models can be used to simulate transient phenomena. There are two types of inertial models: (i) water hammer or elastic models, which consider the elasticity of the pipe and the water, and (ii) mass oscillation or rigid models, which neglect these factors [23,24]. In this study, the authors developed a mathematical model based on the mass oscillation equation, which is well-suited for predicting water leakages while accounting for system inertia from regulating valves. It is paramount to emphasize the need for more inertial models in the current literature for predicting water leakages in water distribution systems. The authors are pioneering the implementation of this approach for single installations. The proposed model neglects effects related to suspended particles [25] and biofilm formation [26].

This ongoing research aims to develop more advanced models that consider system inertia in more complex water installations to analyze water leakages. Initially, a general formulation applicable to all parallel pipes is demonstrated based on the mass oscillation equation, pressure-head-dependent and pressure-head-independent consumption, and the Swamee–Jain equation. Subsequently, a practical application involving two parallel pipes is presented as an example of how the developed model can be employed.

2. Mathematical Model

2.1. Assumptions

Parallel hydraulic installations are a set of pipes that start at an initial node and end at a common node. It is implied that water flows in all parallel pipes reach an end node, and then a standard pressure head is obtained over time. In addition, water leakages should be considered within a water balance at the common node. The total flow rate at the common node must be computed as the sum of water flows and leakages passing in all parallel pipes. Parallel pipes can have different lengths, internal pipe diameters, absolute roughness values, opening/closing maneuvers in regulating valves, and accessories. Parallel systems have a maximum of four pipes, but two are usually presented in most hydraulic installations.

The numerical resolution of a set of algebraic-differential governing equations for simulating water leakages in parallel pipelines considering variations of the system inertia is developed in this section. An analytical solution to the equations is not available; therefore, the numerical resolution is conducted using the solver ODE23 of the Simulink library in Matlab, which is used for solving nonstiff differential equations involving two simultaneous single-step formulations (second and third order) [27]. Boundary and initial conditions are presented, and parallel water pipeline examples are solved. The continuity and momentum equations form a pair of quasi-linear ordinary differential equations regarding each parallel pipe's water flow rate and pressure head.

The proposed model has the assumptions as follows:

- Water movement is simulated using the mass oscillation equation [23,24];
- Friction losses are calculated based on the absolute roughness, internal pipe diameter, and Reynolds number, a common practice in the current literature [28];
- Minor losses are computed using a dimensionless coefficient;
- A resistance coefficient is used for evaluating valve losses in a hydraulic grade line.

2.2. Governing Equations

This section presents the governing equations to study the problem of computing water leakage volume in parallel pipe systems considering maneuvers in regulating valves. Figure 1 presents a plan view illustrating a general schematic of multiple parallel pipes. Each pipe within this configuration possesses unique characteristics, including its internal diameter, absolute roughness, length, minor loss coefficient, and a resistance coefficient associated with a regulating valve.

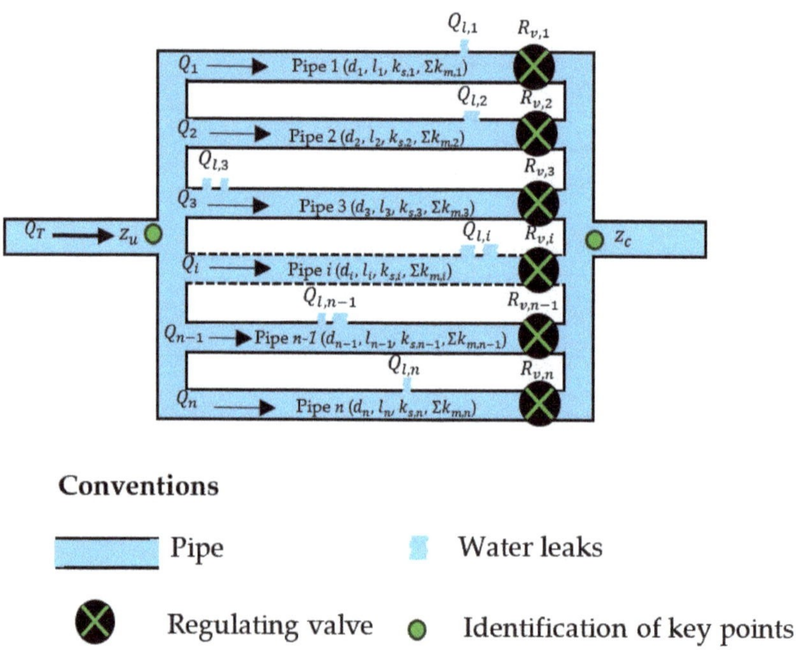

Figure 1. Plan view illustrating a general schematic of n parallel pipes.

a. Mass oscillation equation

This equation considers the system inertia, which must be considered when opening/closing maneuvers in regulating valves are acting. The equation neglects pipe and water elasticity since water leakages tend to reduce pressure surges.

For the total parallel pipes (n), the following equation follows the behavior of water movement [24]:

$$\frac{dQ_i}{dt} = \frac{\left[z_u - z_c - \frac{p_c}{\gamma_w}\right] g\pi d_i^2}{4l_i} - \frac{2f_i Q_i|Q_i|}{\pi d_i^3} - \frac{R_{v,i} Q_i |Q_{in}| g\pi d_i^2}{4l_i} - \frac{2\sum k_{m,i} Q_i |Q_i|}{\pi l_i d_i^2} \quad (1)$$

where z_u = pipe elevation in the first node, z_c = pipe elevation in the common node, p_c = pressure in the common node, γ_w = water unit weight, f = friction factor, Q = water flow rate, d = internal pipe diameter, l = pipe length, $\sum k_m$ = minor losses coefficient, R_v = resistance coefficient of a regulating valve, and g = gravitacional aceleration. The subscript i refers to an analazed parallel pipe.

b. Continuity equation

The continuity equation should be established, considering the water balance guidelines recommended by the International Water Association (IWA) or the national regulations endorsed by local government authorities. In all cases, the water balance should encompass, at the very least, factors such as domestic consumption (Q_d), commercial consumption (Q_c), industrial consumption (Q_i), official consumption (Q_o), physical leakages (Q_l), illegal use (Q_{ie}), and flow by metering error (Q_m), among other variables.

To establish the water balance, it is crucial to consider both pressure-head-dependent (PHDC) and pressure-head-independent (PHIC) consumption.

PHIC can be modeled based on a coefficient modulation (C) over time and the mean value of an analyzed consumption. For instance, domestic consumption can be simulated with the expression $Q_d = C_d\overline{Q_d}$, where C_d = coefficient modulation for domestic consumption, and $\overline{Q_d}$ = mean value of Q_d. Similarly, the remaining PHICs can be modeled.

Regarding the PHDC, only physical leakages will be considered because water utilities are grappling with significant issues concerning the quantity of water lost from water distribution systems. For a parallel pipe, the water leakages can be calculated as [18]:

$$Q_{l,i} = K_{f,i}\left(\frac{p_c}{\gamma_w}\right)^{\beta} \quad (2)$$

where K_f = emitter coefficient and β = exponent coefficient.

The exponent coefficient can be varied from 0.5 to 2.5. A value of 0.5 corresponds to a typical behavior found in many hydraulic installations. In addition, the exponent coefficient (β) must be the same for all analyzed pipes. In Equation (2), the n parallel pipes maintain a consistent average pressure across the lines, corresponding to the shared node's pressure head.

The proposed model is built upon a prior calibration of the emitter coefficient and exponent coefficient in Equation (2) for an extended period simulation.

To establish the water balance, it is essential to considered that the n pipes must transport both PHDCs and PHICs. In this case, only domestic, industrial, commercial, and official consumption and physical leakages in all n parallel pipes are considered for the water balance. Consequently, the continuity equation is as follows:

$$Q_1 + Q_2 + \cdots + Q_i + \cdots + Q_n = C_d\overline{Q_d} + C_i\overline{Q_i} + C_c\overline{Q_c} + C_o\overline{Q_o} + K_{f,1}\left(\frac{p_c}{\gamma_w}\right)^{\beta} + K_{f,2}\left(\frac{p_c}{\gamma_w}\right)^{\beta} + \cdots + K_{f,i}\left(\frac{p_c}{\gamma_w}\right)^{\beta} + \cdots + K_{f,n}\left(\frac{p_c}{\gamma_w}\right)^{\beta} \quad (3)$$

Rearranging the terms in Equation (3) in function of the pressure head in the common node, Equation (4) is established:

$$\frac{p_c}{\gamma_w} = \left(\frac{Q_1 + Q_2 + \cdots + Q_i + \cdots + Q_n - C_d\overline{Q_d} - C_i\overline{Q_i} - C_c\overline{Q_c} - C_o\overline{Q_o}}{K_{f,1} + K_{f,2} + \cdots + K_{f,i} + \cdots + K_{f,n}}\right)^{\frac{1}{\beta}} \quad (4)$$

c. Friction factor equation

The Swamee–Jain equation has been used to compute the friction factor. The friction factor in the turbulence zone varies as a function of absolute roughness, internal pipe diameter, and Reynolds number. This equation has been extensively applied within the Reynolds number range of 3×10^3 to 3×10^8 and absolute roughness (k_s) ranging from 10^{-6} to 2×10^{-2} mm. It encompasses a wide range of important pipe materials, including Polyvinyl Chloride (k_s = 0.0015 mm), Concrete Cylinder Pipe (k_s = 0.12 mm), and High-density Polyethylene (k_s = 0.007 mm), among others. The Swamee–Jain equation is presented as follows:

$$f_i = \frac{0.25}{\left[\log\left(\frac{k_{s,i}}{3.7d_i} + \frac{5.74}{\mathrm{Re}_i^{0.9}}\right)\right]^2} \tag{5}$$

where Re = Reynolds number ($\mathrm{Re} = \frac{v_i d_i}{\upsilon}$). Here, υ = kinematic viscosity and v_i = water velocity of a parallel pipe.

2.3. Numerical Resolution

To solve the system of algebraic differential equations, it is necessary to determine the number of equations and corresponding variables. This set of formulations comprises Equations (1), (4) and (5), which must be applied to the i parallel pipes. The variables in this problem are Q_i, p_c/γ_w, and f_i. The numerical resolution involves a total of $2i + 1$ equations, all with the same number of variables. This implies that analyzing water leakages in a parallel pipe system and considering maneuvers in a regulating valve has a unique solution. The numerical resolution is carried out using the Simulink library in Matlab. Table 1 lists the total number of variables that need to be computed over time depending on the number of parallel pipes in an analyzed system. Based on the information reported in Table 1, if $i = 1$, then the mathematical model is used to simulate the case of a single pipeline, which was the first solution conducted by the authors.

Table 1. Total variables depending on the number of parallel pipes.

i	Variables	Total Variables
1 (single pipeline)	Q_1, f_1, and p_c/γ_w	3
2	Q_1, f_1, Q_2, f_2, and p_c/γ_w	5
3	Q_1, f_1, Q_2, f_2, Q_3, f_3, and p_c/γ_w	7
...	...	...
n	Q_1, f_1, Q_2, f_2, Q_3, f_3,..., Q_n, f_n, and p_c/γ_w	$2n + 1$

Since the mathematical model developed by the authors can be used to quantify the total water leakage volume for maneuvers in regulating valves, it is necessary to establish the initial condition given during the extended period simulation in order to know the initial water flow rate in all parallel pipes ($Q_1(0)$, $Q_2(0)$,..., $Q_n(0)$). The emitter (K_f) and exponent (β) coefficients must be computed during an extended period simulation that requires the calibration of these parameters.

3. Practical Application and Data

3.1. Application to the Case of Two Parallel Pipes

The case of two parallel (n = 2) pipes is considered in this research, as shown in Figure 2. More complex installations can be analyzed with the mathematical model since this model is entirely general.

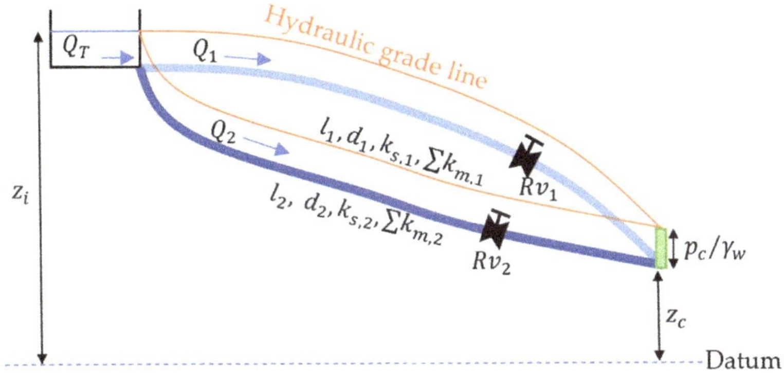

Figure 2. Scheme for two parallel pipes.

The corresponding formulations can be written as shown in Table 2.

Table 2. Algebraic-differential equation system applied to the case of two parallel pipes.

Description	Formula	Equation No.
Mass oscillation equation applied to the first parallel pipe	$\frac{dQ_1}{dt} = \frac{\left[z_u - z_c - \frac{p_c}{\gamma_w}\right] g\pi d_1^2}{4l_1} - \frac{2f_1 Q_1 \lvert Q_1 \rvert}{\pi d_1^3} - \frac{R_{v,1} Q_1 \lvert Q_1 \rvert g\pi d_1^2}{4l_1} - \frac{2\sum k_{m,1} Q_1 \lvert Q_1 \rvert}{\pi l_1 d_1^2}$	(6)
Mass oscillation equation applied to the second parallel pipe	$\frac{dQ_2}{dt} = \frac{\left[z_u - z_c - \frac{p_c}{\gamma_w}\right] g\pi d_2^2}{4l_2} - \frac{2f_2 Q_2 \lvert Q_2 \rvert}{\pi d_2^3} - \frac{R_{v,2} Q_2 \lvert Q_2 \rvert g\pi d_2^2}{4l_2} - \frac{2\sum k_{m,2} Q_2 \lvert Q_2 \rvert}{\pi l_2 d_2^2}$	(7)
Continuity equation	$\frac{p_c}{\gamma_w} = \left(\frac{Q_1 + Q_2 - C_d\overline{Q_d} - C_i\overline{Q_i} - C_c\overline{Q_c} - C_o\overline{Q_o}}{K_{f,1} + K_{f,2}}\right)^{\frac{1}{\beta}}$	(8)
Swamee–Jain equation applied to the first parallel pipe	$f_1 = \frac{0.25}{\left[\log\left(\frac{k_{s,1}}{3.7d_1} + \frac{5.74}{\mathbf{Re}_1^{0.9}}\right)\right]^2}$	(9)
Swamee–Jain equation applied to the second parallel pipe	$f_2 = \frac{0.25}{\left[\log\left(\frac{k_{s,2}}{3.7d_2} + \frac{5.74}{\mathbf{Re}_2^{0.9}}\right)\right]^2}$	(10)

This problem involves a set of five equations with initial conditions given as $Q_1(0) = Q_{1,0}$ and $Q_2(0) = Q_{2,0}$. It constitutes a well-posed problem, the solution of which describes the system's behavior. $Q_{1,0}$ and $Q_{2,0}$ are computed through an extended period simulation. The objective is to calculate the five unknown variables: Q_1, Q_2, f_1, f_2, and p_c/γ_w. The system's behavior is determined through the numerical resolution of Equations (6)–(10). The problem's boundary conditions are defined by a reservoir located upstream and a regulating valve downstream, representing the classical boundary conditions.

3.2. Data

An analysis using an extended period simulation of a two-parallel pipes system revealed that the initial flow rate conditions were $Q_1(0) = 0.078$ m^3/s and $Q_2(0) = 0.045$ m^3/s, similar to the scheme depicted in Figure 2. The characteristics of the two parallel pipes are as follows: $d_1 = 250$ mm, $l_1 = 1200$ m, $d_2 = 200$ mm, $l_2 = 1100$ m, $\sum k_{m,1} = 3$, $\sum k_{m,2} = 5$, $R_{v,1} = 240$ ms^2/m^6, $R_{v,2} = 190$ ms^2/m^6, and $k_{s,1} = k_{s,2} = 0.0015$ mm. The pipe elevation at the first node (z_i) is 35 m, while at the common node (z_c) it is 0 m. The mean values of water flow rate for domestic ($\overline{Q_d}$) and official consumption ($\overline{Q_o}$) are 65 and 22 L/s, respectively. A kinematic viscosity of 1×10^{-6} m/s^2 was considered for the analysis. The remaining consumption Q_i and Q_c are null in this application. Figure 3 shows consumption patterns for both domestic and official purposes throughout the first 180 s. The calibration process, conducted during an extended period simulation, yielded emitter coefficients of $K_{f,1} = 0.0085$ m^3/s/m$^{0.5}$ and $K_{f,2} = 0.0045$ m^3/s/m$^{0.5}$, with an exponent coefficient β of 0.5. The regulating valve is initially opened, followed by a partial closure, and it ultimately returns to its initial position in terms of the percentage of opening. During this time, modulation coefficients of 0.3 and 0.4 are observed for domestic and official consumption, respectively.

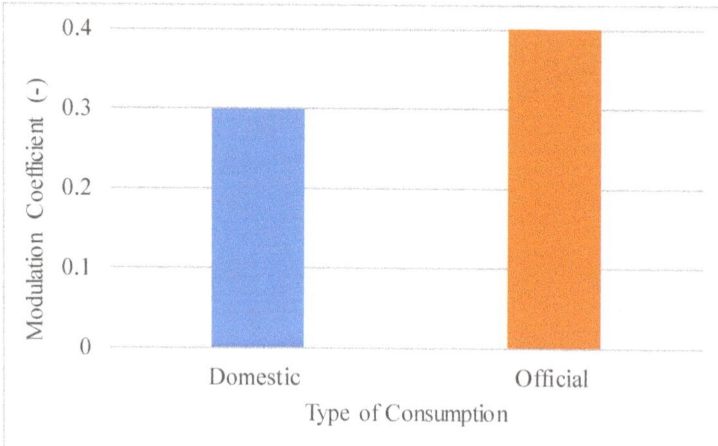

Figure 3. Domestic and official consumption patterns for the first 180 s.

4. Results

The numerical resolution of the five unknown variables (Q_1, Q_2, $\frac{p_c}{\gamma_w}$, f_1, and f_2) was accomplished by analyzing data from the two parallel pipes, as depicted in Figure 4. The simulation was conducted for the initial 180 s.

The analysis of these five variables was conducted. Figure 4a shows the water flow rate patterns for both pipes. The minimum value is 45.1 L/s (at 13.0 s) for pipe 1, while pipe 2 reaches a minimum value of 35.9 L/s (at 4.5 s). After t = 45 s, both pipes exhibit nearly identical trends, with constant values of 60.6 L/s for pipe 1 and 37.5 L/s for pipe 2. However, the variable Q_2 displays some fluctuations over time, reaching a second peak value of 41.4 m at 30.0 s. The pressure head pulses at the common node are presented in Figure 4b. The initial pressure head value is 53.06 m. Rapidly, the partial closure of both regulating valves causes the pressure head to decrease to a minimum value of 17.41 m at 12.0 s. Subsequently, the pressure head increases and stabilizes at around 28.75 m (at 60.0 s). Figure 4c illustrates the friction factor for both parallel pipes, with values varying from 0.0137 to 0.0152 for pipe 1 and from 0.0146 to 0.0152 for pipe 2.

The total flow rate (Q_T) and flow rate of leaks (Q_{Tl}) were computed for the two parallel pipes. These two formulations are used to elucidate the extent of water volume leakage during transient events involving regulating valves. The results of these variables are illustrated in Figure 5. It becomes evident that a higher emitter coefficient corresponds to an increased water volume leakage, as exemplified by pipes 1 and 2. Furthermore, the total flow rate ($Q_1 + Q_2$) exhibits a comparable pattern to the total flow rate of leaks (Q_{Tl}). The occurrences of maximum and minimum water flow coincide with the extreme values of water leakages. For example, the minimum values for both total water flow and water flow from leaks are 82.54 L/s and 52.24 L/s, respectively, occurring at t =12.0 s. For the initial 180 s, the injected water volume to the system is 17.46 m^3, while the water leakage volume is 12.35 m^3.

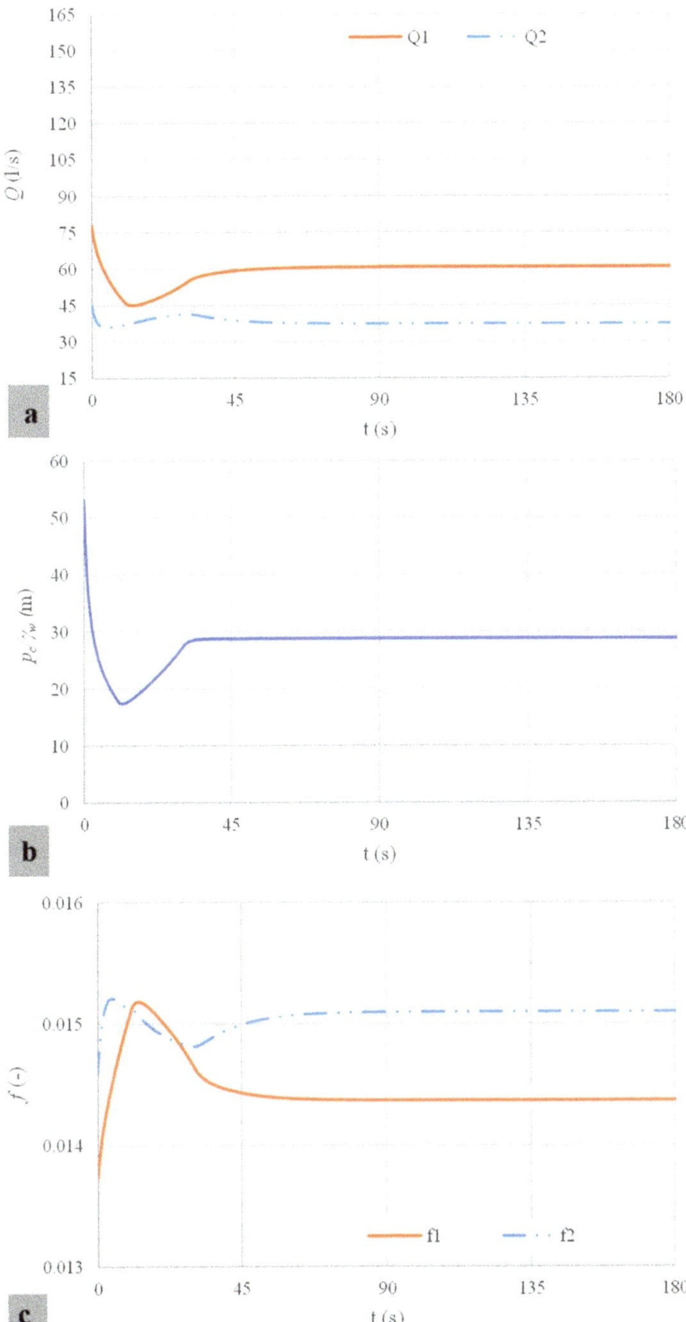

Figure 4. Results of main variables: (**a**) water flow rate; (**b**) pressure head; (**c**) friction factor.

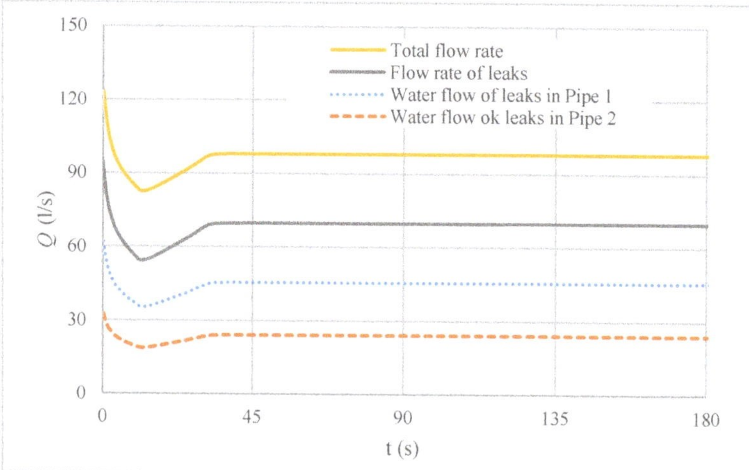

Figure 5. Analysis of water flow and leakages in parallel pipes.

Finally, a comparison was conducted between a single pipe installation and the parallel installations under analysis, as illustrated in Figure 6. The characteristics of the single pipe match those of Pipe 1 in the parallel system. It was observed that higher pressure head values at the end of the system were achieved with an increased number of parallel pipes. Consequently, the parallel pipe configuration led to a greater quantity of water leakages when compared to a single installation. In this analysis, the total water leakage during the first 180 s in the single pipe amounted to 7.57 m^3, which was lower than the value obtained in the parallel system (12.35 m^3).

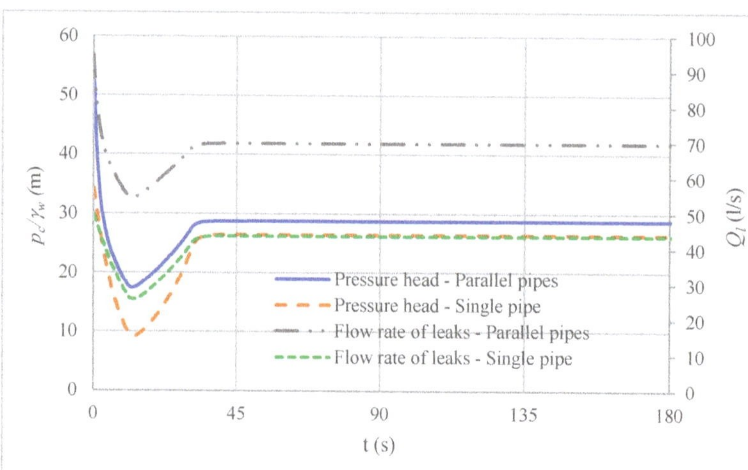

Figure 6. Comparison between single and parallel systems.

5. Discussion

The proposed model is well-suited for assessing transient events occurring within a timeframe, given its consideration of system inertia [23]. The term dQ/dt is employed to account for valve maneuvers. Utilities typically utilize extended period simulation to implement strategies for addressing water leakages in water distribution systems [9]. This is crucial as it enables more accurate computation.

Detecting leaks in water distribution systems poses a challenge for water utilities. Nevertheless, once leaks have been identified, the proposed model can be employed to assess water volume losses under various valve maneuver scenarios. Neither the extended period simulation (which serves as the current leak prediction model) [16] nor the proposed model (which incorporates system inertia for enhanced accuracy) can pinpoint the exact locations of leaks.

The PDE system, or elastic model equations, represents the more intricate formulations for transient events, entailing the analysis of water and pipe elasticity. However, suppose water cannot compress or expand during a transient event or the pipe cannot deform its thickness. In that case, system inertia can be examined using a rigid column approach, as proposed by our model [29]. Leaks operate similarly to water-release orifices; consequently, they reduce head pressure.

A mathematical model for the accurate determination of water volume leakages in all parallel pipe systems, which is based on the mass oscillation equation (rigid water column model), has been presented in this research. It signifies a novel advancement in the current literature, as no governing equations have been previously documented for modeling this situation. The proposed model permits the computation of water leakages even when regulating valves are acting, which implies that the system inertia is considered in calculations. In this sense, the mathematical model is more robust than the extended period simulation in this scenario. The mathematical model is presented in a general way and can be used to analyze parallel pipe systems. The numerical resolution includes the mass oscillation equation to know the behavior of water movement, the analysis of pressure-head-dependent and pressure-head-independent consumption, and the friction factor computation.

Based on the results of the practical application, some remarks can be drawn:

- The practical application is composed of two parallel pipes, and its results can be used for engineers and designers to compute the total leakage volume when two regulating valves are operated following different maneuvers. The mathematical model shows that each problem of parallel pipe requires $2i + 1$ equations to be solved. In this case ($i = 2$), the variables to compute are Q_1, Q_2, $\frac{p_c}{\gamma_w}$, f_1, and f_2. Due to the regulating valves acting, the pressure head patterns are changing even for constant consumption (domestic and official), ranging from 53.06 m to 28.75 m. The two parallel pipes' water flow and friction factor patterns are different and should be separately computed.
- When more parallel pipes are added to reinforce a line, the amount of water volume of leakages is increased since, during the operational time, water installations have more orifices along their length, provoking more water losses. It is then of the utmost importance to reduce the number of parallel pipes during the design stage to save energy production in the injected water flow.

The primary limitation of the proposed model is that it has not been specifically designed for complex installations. The current research serves as an initial phase, laying the groundwork for extending this methodology to more intricate networks. Additionally, it is noteworthy that the numerical resolution of the proposed model is more intricate compared to the extended period simulation, resulting in a longer computational time.

In summary, the proposed model can calculate water leaks in much greater detail than the quasi-static models that are usually used. Thus, the rigid inertial model is suitable for detecting rapid flow variations (for example, rapid regulating valve maneuvers) with greater precision.

Author Contributions: Conceptualization, V.S.F.-M., A.A.-P. and O.E.C.-H.; methodology, A.A.-P. and O.E.C.-H.; formal analysis, O.E.C.-H.; validation, A.A.-P., and V.S.F.-M.; writing—original draft preparation, A.A.-P. and O.E.C.-H.; writing—review and editing, V.S.F.-M. All authors have read and agreed to the published version of the manuscript.

Funding: No funding from either external or internal sources was secured for this research project.

Data Availability Statement: Databases are available from the corresponding author.

Conflicts of Interest: The authors declare no conflicts of interest.

Abbreviations

The following abbreviations were used in this research:

C:	coefficient modulation $(-)$
d:	internal pipe diameter (m)
f:	friction factor (-)
K_f:	emitter coefficient ($m^3/s/m^{0.5}$)
k_s:	absolute roughness (m)
g:	gravitational acceleration (m/s^2)
i:	a specified pipe (-)
l:	pipe length (m)
n:	number of parallel pipes (-)
t:	time (s)
p_c:	pressure in the common node (Pa)
R_v :	resistance coefficient of a regulating valve (ms^2/m^6)
Re:	Reynolds number (-)
Q:	water flow rate (m^3/s)
Q_c:	commercial consumption (m^3/s)
Q_d:	domestic consumption (m^3/s)
Q_i:	industrial consumption (m^3/s)
Q_o:	official consumption (m^3/s)
Q_l:	physical leakages (m^3/s)
Q_{ie}:	illegal use (m^3/s)
Q_m:	flow by metering error (m^3/s)
v:	water velocity in a pipe (m/s)
z_i:	pipe elevation in the first node (m)
z_c:	pipe elevation in the common node (m)
υ:	kinematic viscosity (m/s^2)
β:	exponent coefficient of leakages (-)
$\sum k_m$:	minor losses coefficient (-)
γ_w:	water unit weight (N/m^3)

References

1. Lambert, A.; Hirner, W. *Losses from Water Supply Systems: A Standard Terminology and Recommended Performance Measures*; IWA: London, UK, 2000.
2. Lambert, A.O.; Brown, T.G.; Takizawa, M.; Weimer, D. A review of performance indicators for real losses from water supply systems. *J. Water Supply Res. Technol. AQUA* **1999**, *48*, 227–237. [CrossRef]
3. IBNET. IBNET Indicator Heat Map. February 2021. Available online: https://database.ib-net.org/Reports/Indicators/HeatMap (accessed on 1 February 2024).
4. United Nations. Sustainable Development Goals. Available online: https://sdgs.un.org/goals (accessed on 1 February 2024).
5. Bibri, S.E.; Krogstie, J. Smart sustainable cities of the future: An extensive interdisciplinary literature review. *Sustain. Cities Soc.* **2017**, *31*, 183–212. [CrossRef]
6. Ishiwatari, Y.; Mishima, I.; Utsuno, N.; Fujita, M. Diagnosis of the ageing of water pipe systems by water quality and structure of iron corrosion in supplied water. *Water Sci. Technol. Water Supply* **2013**, *13*, 178–183. [CrossRef]
7. Li, R.; Huang, H.; Xin, K.; Tao, T. A review of methods for burst/leakage detection and location in water distribution systems. *Water Sci. Technol. Water Supply* **2015**, *15*, 429–441. [CrossRef]
8. Cimorelli, L.; D'Aniello, A.; Cozzolino, L.; Pianese, D. Leakage reduction in WDNs through optimal setting of PATs with a derivative-free optimizer. *J. Hydroinform.* **2020**, *22*, 713–724. [CrossRef]
9. Ramos, H.M.; Kuriqi, A.; Besharat, M.; Creaco, E.; Tasca, E.; Coronado-Hernández, O.E.; Pienika, R.; Iglesias-Rey, P. SmartWater Grids and Digital Twin for the Management of System Efficiency in Water Distribution Networks. *Water* **2023**, *15*, 1129. [CrossRef]
10. Alzamora, F.M.; Carot, M.H.; Carles, J.; Campos, A. Development and Use of a Digital Twin for the Water Supply and Distribution Network of Valencia (Spain). In Proceedings of the 17th International Computing & Control for the Water Industry Conference, Exeter, UK, 1–4 September 2019.

11. Jiang, Z.; Guo, Y.; Wang, Z. Digital Twin to Improve the virtual-real Integration of Industrial Iot. *J. Ind. Inf. Integr.* **2021**, *22*, 100196. [CrossRef]
12. Liu, M.; Fang, S.; Dong, H.; Xu, C. Review of Digital Twin about Concepts, Technologies, and Industrial Applications. *J. Manuf. Syst.* **2021**, *58*, 346–361. [CrossRef]
13. Niu, W.J.; Feng, Z.K. Evaluating the performances of several artificial intelligence methods in forecasting daily streamflow time series for sustainable water resources management. *Sustain. Cities Soc.* **2021**, *64*, 102562. [CrossRef]
14. Lu, Q.; Parlikad, A.K.; Woodall, P.; Don Ranasinghe, G.; Xie, X.; Liang, Z.; Konstantinou, E.; Heaton, J.; Schooling, J. Developing a Digital Twin at Building and City Levels: Case Study of West Cambridge Campus. *J. Manag. Eng.* **2020**, *36*, 05020004. [CrossRef]
15. Asadi, S.; Nazari-Heris, M.; Rezaei Nasab, S.; Torabi, H.; Sharifironizi, M. An Updated Review on Net-Zero Energy and Water Buildings: Design and Operation. In *Food-Energy-Water Nexus Resilience and Sustainable Development: Decision-Making Methods, Planning, and Trade-Off Analysis*; Springer: Cham, Switzerland, 2020; pp. 267–290.
16. Ávila, C.; Sánchez-Romero, F.-J.; López-Jiménez, P.A.; Pérez- Sánchez, M. Improve Leakage Management to Reach Sustainable Water Supply Networks through by Green Energy Systems. Optimized Case Study. *Sustain. Cities Soc.* **2022**, *83*, 103994. [CrossRef]
17. Jowitt, P.W.; Xu, C. Optimal valve control in water-distribution networks. *J. Water Resour. Plan. Manag.* **1990**, *116*, 455–472. [CrossRef]
18. Giustolisi, O.; Savic, D.; Kapelan, Z. Pressure-driven demand and leakage simulation for water distribution networks. *J. Hydraul. Eng.* **2008**, *134*, 626–635. [CrossRef]
19. Al-Washali, T.; Sharma, S.; Kennedy, M. Methods of assessment of water losses in water supply systems: A review. *Water Resour. Manag.* **2016**, *30*, 4985–5001. [CrossRef]
20. Samir, N.; Kansoh, R.; Elbarki, W.; Fleifle, A. Pressure control for minimizing leakage in water distribution systems. *Alex. Eng. J.* **2017**, *56*, 601–612. [CrossRef]
21. Schwaller, J.; Van Zyl, J.V. Modeling the pressure-leakage response of water distribution systems based on individual leak behavior. *J. Hydraul. Eng.* **2015**, *141*, 04014089. [CrossRef]
22. Coronado-Hernández, O.E.; Pérez-Sánchez, M.; Arrieta-Pastrana, A.; Fuertes-Miquel, V.S.; Coronado-Hernández, J.R.; Quiñones-Bolaños, E.; Ramos, H.M. Dynamic effects of a regulating valve in the assessment of water leakages in single pipelines. *Water Resour. Manag.* **2024**. [CrossRef]
23. Fuertes-Miquel, V.S.; Coronado-Hernández, O.E.; Mora-Meliá, D.; Iglesias-Rey, P.L. Hydraulic modeling during filling and emptying processes in pressurized pipelines: A literature review. *Urban Water J.* **2019**, *16*, 299–311. [CrossRef]
24. Coronado-Hernández, O.E.; Fuertes-Miquel, V.S.; Iglesias-Rey, P.L.; Martínez-Solano, F.J. Rigid water column for simulating the emptying process in a pipeline using pressurized air. *J. Hydraul. Eng.* **2018**, *144*, 06018004. [CrossRef]
25. Prest, E.I.; Schaap, P.G.; Besmer, M.D.; Hammes, F. Dynamic Hydraulics in a Drinking Water Distribution System Influence Suspended Particles and Turbidity, but Not Microbiology. *Water* **2021**, *13*, 109. [CrossRef]
26. Tsagkari, E.; Connelly, S.; Liu, Z.; McBride, A.; Sloan, W.T. The role of shear dynamics in biofilm formation. *npj Biofilms Microbiomes* **2022**, *8*, 33. [CrossRef] [PubMed]
27. Eshkabilov, S.L. *Practical MATLAB Modeling with Simulink: Programming and Simulating Ordinary and Partial Differential Equations*; Apress: Berkeley, CA, USA, 2020.
28. Altowayti, W.A.H.; Othman, N.; Tajarudin, H.A.; Al-Dhaqm, A.; Asharuddin, S.M.; Al-Gheethi, A.; Alshalif, A.F.; Salem, A.A.; Din, M.F.M.; Fitriani, N.; et al. Evaluating the Pressure and Loss Behavior in Water Pipes Using Smart Mathematical Modelling. *Water* **2021**, *13*, 3500. [CrossRef]
29. Abreu, J.; Cabrera, E.; Izquierdo, J.; García-Serra, J. Flow Modeling in Pressurized Systmes Revisited. *J. Hydraul. Eng.* **1999**, *125*, 1154–1169. [CrossRef]

water

MDPI

Article

Influence of Entrapped Air on Hydraulic Transients During Rapid Closure of a Valve Located Upstream and Downstream of an Air Pocket in Pressurised Pipes

Oscar Pozos-Estrada

Institute of Engineering, Coordinación de Ingeniería Hidráulica, National Autonomous University of Mexico, Ciudad de Mexico C.P. 04510, Mexico; opozose@iingen.unam.mx

Abstract: This paper examines the dual impact of trapped air on fluid transients in pressurised conduits, highlighting both its beneficial and detrimental impacts. This research analyses transient pressures caused by rapid valve closure in pipelines that contain air pockets and small bubbles dispersed within the liquid phase, by a hydraulic jump occurring at the downstream edge of the pockets. Experiments and numerical simulations were conducted with the valve positioned at the ends of the test section on both the inflow and outflow sides. A numerical model utilising the four-point centred scheme and method of characteristics was developed to resolve the governing equations of two-phase flow and was experimentally validated. The results indicate that entrapped air significantly influences hydraulic transients. When the valve is positioned downstream, air pockets and bubbles reduce pressure transients, illustrating a favourable effect. Conversely, when the valve is positioned upstream, adverse pressure transients occur, highlighting a detrimental impact. These outcomes underscore the importance of considering trapped air in pipeline systems, as its existence can either mitigate or exacerbate transient pressures depending on the configuration of the pipeline. The research highlights the significance of considering entrapped air in the design and evaluation of pressurised conduits to improve performance and prevent adverse effects.

Keywords: hydraulic transients; rapid closure; hydraulic jump; air pocket; air bubbles

1. Introduction

Although it is practically impossible to completely eliminate air from pipelines, identifying and addressing the sources of air ingress can significantly reduce its occurrence. For instance, water under standard conditions contains about 2% dissolved air by volume [1]. Additionally, pipelines are initially filled with air, and incomplete expulsion through valves, vents, and standpipes can leave residual air pockets at high points within the system [2]. Similarly, pumps can introduce air through vortex action during suction, contributing approximately 5–10% of the water discharge [3]. Furthermore, if pipeline pressure falls below atmospheric levels, air may infiltrate through faulty seals at joints and valves [4]. These factors collectively highlight the challenges of managing air in pipeline systems and the importance of targeted measures to minimise its presence.

The trapped air within pipelines may result in a variety of operational challenges. Notably, the formation of air pockets at elevated sections within enclosed conduits can substantially reduce flow rates by increasing head loss significantly [5,6]. Additionally, free air can compromise the accuracy of flow measurement devices. The intermittent shift from partially full to full flow conditions caused by these air pockets can induce vibrations within the pipeline. In extreme cases, substantial accumulation of air may trigger considerable pressure surges, resulting in vibrations and potential structural damage [7]. Moreover, large air pockets have the capacity to partially or completely block water flow, thereby reducing the efficiency and capacity of both pump-driven and gravity-fed systems [8]. Furthermore, when distributed along pipelines, these pockets can decrease overall system capacity and consequently increase energy consumption [9].

Air entrainment within water pipelines can result in the formation of pockets at the highest elevations of these systems, particularly when air valves are either absent or malfunctioning. These trapped air pockets can cause the water flowing beneath them to shift into a free-surface flow regime. This phenomenon has been explored by several researchers [10–12].

The rapid release of air from pipelines via air valves or orifices can trigger substantial pressure transients. Various researchers have investigated this phenomenon, highlighting crucial factors that influence peak transient pressures. These include the dimensions of orifices, characteristics of air pockets, driving heads, and the length of water columns [13–20]. Recent studies by Aguirre-Mendoza et al. [21] have utilised 2D CFD software to model rapidly filling manoeuvres involving air valves. Their work considers diverse air pocket sizes and pressure variations. Additionally, Aguirre-Mendoza et al. [22] examined how uncontrolled filling processes affect air pocket pressure using similar CFD models. This scenario often involves high velocities in water columns that can cause premature closure of air valves, resulting in additional pressure peaks. In a related study, Hurtado-Misal et al. [23] employed a 2D computational fluid dynamics program to analyse sub-atmospheric pressure patterns during the emptying of irregular pipelines without air valves.

In 2022, Fuertes-Miquel [24] underscored the significance of air valves in water distribution systems. Nevertheless, it is acknowledged that installing air valves does not invariably guarantee optimal system performance. Similarly, Ramos et al. [25] investigated the intricate dynamics of two-phase flows within pipelines, both with and without air vents, thereby highlighting the role and limitations of current assumptions regarding air valve functionality. Additionally, Tasca et al. [26] stressed the importance of precise characterisation of air valves for reliable transient simulations. They identified discrepancies in characteristic curves provided by manufacturers and demonstrated how these inaccuracies can result in substantial errors.

The impact of trapped air on pressure surges in water distribution systems is multifaceted, exhibiting both advantageous and disadvantageous effects. These effects are contingent upon several factors, including the volume of air present, its spatial distribution within the pipeline, the geometry of the system itself, and the nature of the transient event triggering these phenomena [27,28]. In water and wastewater pipelines, entrapped air commonly manifests as either stationary pockets or dispersed bubbles suspended within the liquid phase [29]. For instance, research has demonstrated that small air pockets can substantially amplify transient pressures to levels capable of causing pipe rupture [30–34]. Conversely, large air pockets forming within pipelines can serve to alleviate overpressures generated by hydraulic transients [35,36].

The occurrence of small air pockets generating significant overpressures is a consequence of its low density, which leads to minimal inertia and allows for rapid compression. In contrast, larger air pockets act as cushions or dampers, playing a key role in mitigating transient pressures. Gahan [37], following an extensive review of studies on hydraulic transients involving entrapped air, highlights that categorising an air pocket as small or large depends on its influence on transient behaviour. This distinction underscores the importance of understanding how air pocket size affects system dynamics in pipeline design and analysis.

Similarly, the uniform distribution of air in the form of small bubbles adds complexity to predicting its impact on hydraulic systems. A significant consequence is the considerable reduction in wave propagation speed, even when only a small amount of air is present. When a homogeneous mixture of water and air flows through a pressurised conduit, transient pressures are typically lowered due to the internal reflection of pressure surges within water–air mixtures [38]. However, the magnitude of transient pressures is also influenced by the proportion of air in the mixture. In some cases, these pressures can exceed those observed in scenarios where air is not considered, a phenomenon linked to the expansion and compression cycles of air bubbles [39]. This dual behaviour highlights the intricate role of air bubbles in modifying transient responses in pipeline systems. This study builds upon earlier work by Pozos [40] to further investigate numerically and experimentally how trapped air pockets and two-phase mixtures collectively influence hydraulic transients in pipelines.

This paper introduces a numerical model that integrates the four-point centred scheme and the method of characteristics (MOC). The four-point centred scheme is extensively used for simulating unsteady flows in both pressurised and open-channel systems, owing to its computational robustness and reliability [41–43]. Furthermore, numerical models based on the MOC are recognised for yielding accurate results and have been shown to be effective [44]. These models are now actively employed in the development and analysis of water pipelines, particularly in scenarios characterised by cavitation phenomena, the presence of entrapped air, and the identification of leaks in pipelines [45–47]. The combination of these methodologies ensures a robust framework for addressing complex hydraulic transient phenomena.

The numerical model investigates how the influence of valve position—whether situated upstream or downstream relative to an entrapped gas cavity (air pocket)—affects fluid transient behaviour in systems with gas cavities terminating in a rapid flow transition (hydraulic jump). This jump produces gas bubbles, resulting in a dispersed gas–liquid mixture throughout the liquid medium. Hydraulic transients were induced via the abrupt closure of a butterfly valve. The research demonstrates, both experimentally and numerically, that entrapped pockets with their corresponding void fractions substantially reduce pressure transients when the valve is located downstream of the air pocket. This finding is particularly pertinent for designing systems that are more resilient to pressure surges, thereby minimising the potential for costly repairs and operational downtime. Conversely, less favourable pressure transient characteristics were observed when the valve was situated at the upstream boundary of the pipeline.

Previous research has acknowledged the impact of entrapped air on simplified scenarios, such as single horizontal pipes [19,20,48–51]. However, real-world pipeline systems for water distribution and wastewater management are often more complex, with undulating profiles and varying pipe gradients. This research addresses a gap in the existing literature by examining hydraulic transients in more realistic, undulating pipelines where air pockets and hydraulic jumps are common. This prompted the author to design and construct an experimental apparatus to verify the proposed model. The setup incorporates a dedicated segment designed to replicate an elevated section of a pipeline within a topographically varied configuration. Notably, numerical predictions and experimental results demonstrated strong agreement, underscoring the robustness and precision of the proposed model in capturing the dynamics of hydraulic transients in complex pipeline systems.

This research offers experimental validation for a numerical model designed to simulate hydraulic transients in pipelines containing entrapped air. This validation is of paramount importance, as it ensures that the model can be reliably employed for application in the development and evaluation of real-world piping networks. A comprehensive understanding of air behaviour within pipelines is essential for the effective design, evaluation, and operation of pressurised conduits. By accurately accounting for entrapped air, engineers can enhance system performance and mitigate potential adverse effects, such as pipeline damage or operational inefficiencies. The data presented in this study demonstrates the model's capacity to predict the impact of air pockets on transient pressures under varying operating conditions, thereby providing valuable insights for pipeline design and management.

2. Numerical Model

A numerical model was developed to investigate both the adverse and advantageous impacts of trapped air on hydraulic transients resulting from the sudden closure of a valve in conduits incorporating an entrapped gas cavity and dispersed gas bubbles in the bulk fluid medium. The fundamental partial differential equations governing time-dependent gas–liquid flow regimes were solved employing the four-point centred scheme and the MOC.

This research studies a static gas cavity positioned at the highest point of an experimental setup designed to simulate an undulating pipeline, treating it as a boundary condition. The effects of air pockets on fluid transient behaviour are incorporated into the numerical model using the method of characteristics, alongside methodologies outlined by Chaudhry [52] and Wylie et al. [53]. Additionally, specific assumptions proposed by Borrows and Qiu [45] are applied in the formulation of the numerical model, as detailed below:

(1) The traditional method of characteristics is employed to formulate a system of ordinary differential equations. The equations are solved numerically by employing a finite difference approach aligned with the characteristic directions, utilising a scheme with first-order accuracy. To mitigate numerical dissipation and dispersion errors, interpolation is deliberately omitted during the integration process.
(2) For the purposes of this analysis, a gas cavity of the specified volume is incorporated at a designated location within the numerical model. Additionally, the gas cavity is considered to occupy a fixed segment of the conduit cross-section remaining stationary throughout the simulation.
(3) The celerity is assumed to be invariant throughout the analysis. Moreover, the analytical model accounts for energy dissipation by incorporating friction and local losses.

To enhance computational efficiency, the predetermined gas cavity volume is situated at the junction of adjoining pipe segments. In this study, the cavity is represented as an accumulator with uniform pressure distribution throughout its volume. The compressibility of water is deemed negligible when compared to that of air. Furthermore, the influence of inertial forces and frictional resistance are disregarded. The behaviour of the gas is modelled under the assumption of adhering to a reversible thermodynamic process described by a polytropic relationship, which can be expressed as:

$$H_A V^{\varepsilon} = C_0 \tag{1}$$

In this context, H_A represents the total hydraulic energy (m), V denotes the volume of the entrapped gas cavity (m^3), and C is a constant determined from the equilibrium conditions of the gas cavity. The exponent governing the polytropic process ε was assumed to be 1.4, a value commonly used for rapid transient events, such as those caused by a rapid valve closing manoeuvre. This assumption is supported by numerous experimental and numerical studies, which have shown that fast transients involving trapped gas cavities are more accurately modelled utilising $\varepsilon = 1.4$ [54–57].

Since Equation (1) is applicable at any given moment in time, the relationship is formulated for the joint [j, $N(j)$ + 1] at each step $t + \Delta t$.

$$\left(H_{P_{j,N(j)+1}} - z + H_b\right) V^{\varepsilon}_{P_{j,N(j)+1}} = C_0 \tag{2}$$

where $H_{P_{j,N(j)+1}}$ represents the location of the piezometric line at section (j, $N(j)$ + 1) at $t + \Delta t$ (m), z is defined as the vertical displacement from the centreline of the pipe to a reference datum (m), H_b denotes the barometric pressure head (m), and $V_{P_{j,N(j)+1}}$ indicates the volume of the air pocket at the end of the time step (m^3).

Figure 1 illustrates a gas cavity situated at (j, $N(j)$ + 1) junction of a pipeline, downstream of which a hydraulic jump is observed. The diagram compares the piezometric line or hydraulic grade line for scenarios both with and without an air pocket. Analysis of the figure reveals that the total head loss in the pipeline section, in the presence of a gas cavity, arises from the combined effects of frictional and local losses within the pipe, the head loss attributable to the gas cavity, and the energy dissipation produced by means of turbulence from the hydraulic jump.

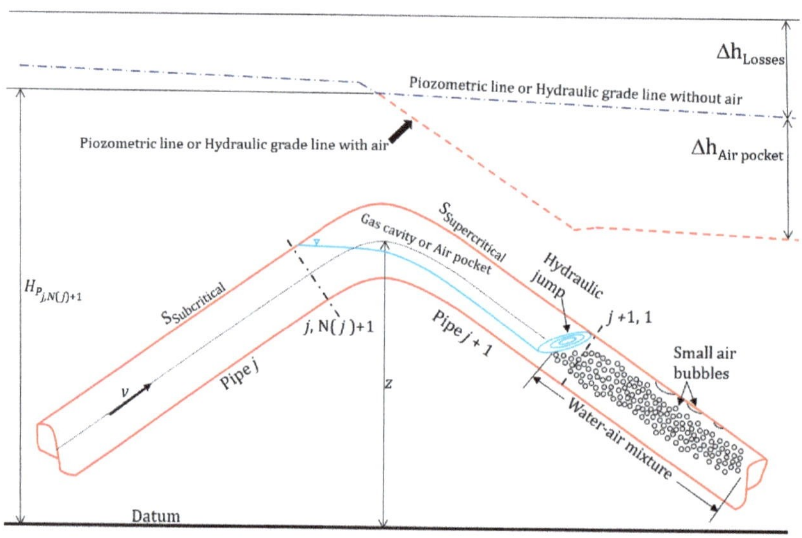

Figure 1. Notation for the air pocket location.

For clarity, the subscript P is utilised to denote variables whose values remain unknown at $t + \Delta t$. Conversely, variables excluding this subscript represent known values at the start of the time step, t.

The application of mass conservation principles to the gas cavity yields a continuity equation as follows:

$$V_{P_{j,N(j)+1}} = V_i + \frac{\Delta t}{2}\left[\left(Q_{P_{j+1,1}} + Q_{j+1,1}\right) - \left(Q_{P_{j,N(j)+1}} + Q_{j,N(j)+1}\right)\right] \tag{3}$$

where V_i represents the initial gas cavity volume at t (m^3). The duration of the time interval is denoted by Δt (s). The liquid discharge rates at the upstream and downstream boundaries of the gas cavity at ($t + \Delta t$) are given by $Q_{P_{j,N(j)+1}}$ and $Q_{P_{N(j)+1,\,1}}$, respectively. Similarly, $Q_{j,N(j)+1}$ and $Q_{j+1,1}$ correspond to the liquid discharge rates at the upstream and downstream boundaries of the gas cavity at the start of the time interval (t).

The domain limits are modelled by means of the MOC, with the forward and backward characteristic equations evaluated upon completion of every discrete time interval, expressed as follows:

$$Q_{P_{j,N(j)+1}} = C_{(+)} - C_{a_j} H_{P_{j,N(j)+1}} \tag{4}$$

$$Q_{P_{N(j)+1,1}} = C_{(-)} + C_{a_{j+1}} H_{P_{N(j)+1,1}} \tag{5}$$

where

$$C_{(+)} = Q_{j,N(j)+1} + \frac{gA_j}{c_j} H_{j,N(j)+1} - \frac{f_j \Delta t_j}{2D_j A_j} Q_{j,N(j)+1} \left| Q_{j,N(j)+1} \right| \tag{6}$$

$$C_{(-)} = Q_{j+1,1} - \frac{gA_{j+1}}{c_{j+1}} H_{j+1,1} - \frac{f_{j+1} \Delta t_{j+1}}{2D_{j+1} A_{j+1}} Q_{j+1,1} \left| Q_{j+1,1} \right| \tag{7}$$

where c denotes the velocity of wave transmission in water devoid of any gases (m/s). Following Wylie et al. [53], Equation (8) was employed within the model to calculate c. The variable A signifies the internal transverse area of the pipe (m^2), D indicates the diameter of the pipe (m), f is the Darcy–Weisbach resistance coefficient, and g represents gravitational acceleration (approximately 9.81 m/s^2).

$$c = \sqrt{\frac{eEE_w}{\rho(Ee + DE_w)}} \tag{8}$$

In this context, E_w denotes the compressibility modulus of water (GPa), E represents the elastic modulus of the material constituting the pipe (GPa), e signifies the wall thickness of the pipe (m), and ρ represents the density of the liquid (kg/m^3).

Assuming negligible head losses at pipeline junctions, then:

$$H_{P_{j,N(j)+1}} = H_{P_{j+1,1}} = H_P \tag{9}$$

At this stage, the system of equations comprises five unknown variables: $H_{P_{j,N(j)+1}}$, $H_{P_{j+1,1}}$, $V_{P_{j,N(j)+1}}$, $Q_{P_{j+1,1}}$, and $Q_{P_{j,N(j)+1}}$. By systematically eliminating the last four unknowns through substitution or other algebraic methods results in

$$\left(H_{P_{j,N(j)+1}} + H_b - z\right)\left[C_{air} + \frac{\Delta t}{2}\left(C_{a_j} + C_{a_{j+1}}\right) H_{P_{j,N(j)+1}}\right]^{\epsilon} = C_0 \tag{10}$$

$$C_{air} = V_i + \frac{\Delta t}{2}\left(Q_{j+1,1} - Q_{j,N(j)+1} + C_{(-)} - C_{(+)}\right) \tag{11}$$

To solve for $H_{P_{j,N(j)+1}}$ in Equation (10), it is necessary to employ an iterative technique, such as the Newton-Raphson method. Once $H_{P_{j,N(j)+1}}$ has been determined, the remaining unknown variables can then be calculated using Equations (2)–(9).

To ensure numerical stability and achieve accurate results, the Courant condition was strictly adhered to throughout all pressure transient simulations conducted using the finite difference scheme

$$\frac{\Delta x}{\Delta t} \geq c \tag{12}$$

It is crucial to highlight the analysis of surge pressures in flows comprising two distinct components presents a greater complexity and challenge compared to single-component flows. Moreover, the application of the method of characteristics is often limited by the nonlinear effects that cause compression waves to become increasingly sharp. This limitation arises from the necessity to implement shock-governed analytical constraints at internal solution domain interfaces during characteristic-based flow modelling [58]. Consequently, a 1D homogeneous modelling approach is utilised to investigate transient pressure fluctuations in bubbly flows comprising water and air, occurring at the downstream end of trapped air cavities.

In this study of transient two-component bubbly flow within pipelines, a set of partial differential equations may be derived through the application of a control-volume approach, as initially proposed by Yadigaroglu and Leahy [59]. This method involves developing continuity equations for both phases. Notably, this formulation does not adhere to a strictly separated flow model, adopting the assumption of minimal relative motion between the fluid phases. As a result, the continuity equation governing the liquid component may be represented in terms of its mass conservation over time and space in the following manner:

$$\frac{\partial}{\partial t}[\rho(1-\alpha)A] + \frac{\partial}{\partial x}[\rho(1-\alpha)Av] = 0 \tag{13}$$

where v denotes the velocity of water within the pipe. It is assumed that this velocity is equivalent to that of the gas phase (m/s). Additionally, α represents the air void fraction.

The mass conservation principle applied to the gaseous component may be written in the following form:

$$\frac{\partial}{\partial t}(\rho_g \alpha A) + \frac{\partial}{\partial x}(\rho_g \alpha A v_g) = 0 \tag{14}$$

where ρ_g denotes the density of the gas (kg/m), v_g represents the velocity of the gas phase (m/s), t denotes time (s), and x indicates the longitudinal position along the pipe (m).

The momentum balance for the mixture can be derived using a control volume approach, which excludes the momentum contribution of the gas phase.

$$\frac{\partial}{\partial t}[\rho(1-\alpha)Av_{mix}] + \frac{\partial}{\partial x}\left[\rho(1-\alpha)Av_{mix}^2\right] + A\frac{\partial p}{\partial x} + \pi D\tau_0 - [g\rho(1-\alpha)Asin\phi] = 0 \tag{15}$$

where p represents the pressure at a specified cross-section (Pa). The mixture velocity is denoted by v_{mix} (m/s), which takes up that both water and air travel at equal velocity. τ_o signifies the boundary shear stress (Pa), which can be determined using the Darcy–Weisbach resistance coefficient f, and ϕ denotes the angle at which the pipe is inclined, measured from the horizontal plane.

$$\tau_0 = \frac{f}{8}(1-\alpha)\rho v_{mix}|v_{mix}| \tag{16}$$

The density of most water–gas mixtures, denoted as ρ_{mix} (kg/m^3), can be reasonably estimated using the following approximation:

$$\rho_{mix} = (1-\alpha)\rho \tag{17}$$

The equations mentioned above can be reformulated by incorporating the compressibility moduli of water and air, as well as the elastic modulus of the material constituting the pipe. This reformulation assumes that there is no production or absorption of gas within the system. Following this approach, Equations (13)–(15) can be expressed in a form similar to that presented by Wigger et al. [41]:

$$G_1 = \frac{\partial p}{\partial t} + v\frac{\partial p}{\partial x} - C_1 C_2 \frac{\partial v}{\partial x} = 0 \tag{18}$$

$$G_2 = \frac{\partial p}{\partial t} + v\frac{\partial p}{\partial x} - C_2\left(\frac{\partial \alpha}{\partial t} + v\frac{\partial \alpha}{\partial x}\right) = 0 \tag{19}$$

$$G_3 = \frac{\partial v}{\partial t} + v\frac{\partial v}{\partial x} + C_3\frac{\partial p}{\partial x} = b \tag{20}$$

The coefficients C_1, C_2, C_3 and b which account for elastic properties and the force terms, are detailed in the Appendix A provided at the end of this paper.

Equations (18)–(20) are solved for pressure p and velocity v using a four-point centred finite difference scheme. This method is widely used for simulating unsteady flows in both pressurised and open-channel systems due to its robustness [41–43]. The continuous spatiotemporal domain is discretised using a rectangular lattice structure illustrated in Figure 2. Each grid point is identified by its spatial index (m) and temporal index (n). Importantly, the spatial spacing Δx and temporal spacing Δt between these points do not necessarily have to be uniform [60].

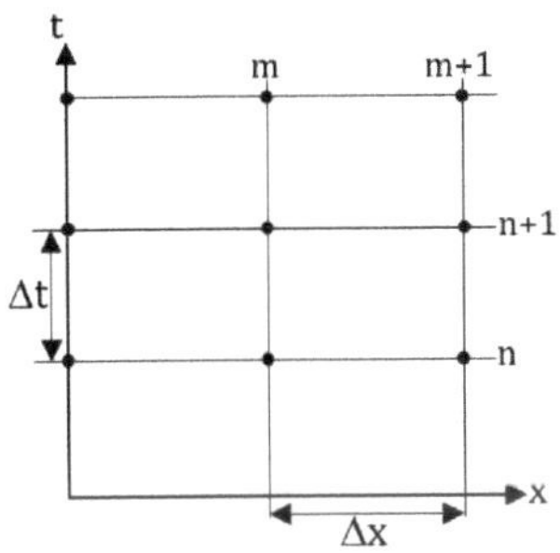

Figure 2. Definition of finite difference grid.

Implicit four-point difference schemes utilise a finite difference method to discretise time derivatives. A forward difference operator is applied to approximate the time derivative term, with the approximation point situated at the midpoint between grid points m and $m+1$ along the x-axis. This approach enhances the accuracy of capturing temporal changes in the solution. For example, when approximating partial derivatives for pressure (p), it can be represented as follows:

$$\frac{\partial p}{\partial t} = \frac{1}{2\Delta t}\left[p_m^{n+1} + p_{m+1}^{n+1} - p_m^n - p_{m+1}^n\right] \tag{21}$$

Similarly, a forward difference operator is employed to approximate spatial derivatives. This operator is applied across two consecutive time steps, incorporating weighting factors θ and $(1-\theta)$ to account for temporal variations.

$$\frac{\partial p}{\partial \mathrm{x}} = \frac{1}{\Delta x}\left[\theta\left(p_{m+1}^{n+1} - p_m^{n+1}\right) + (1-\theta)\left(p_{m+1}^n - p_m^n\right)\right] \tag{22}$$

Equivalent relationships can be established for the derivatives of velocity. Additionally, coefficients, variables, and functions—excluding derivatives—are discretised in time at the specific temporal level where spatial derivatives are computed. This temporal discretisation employs weighting factors akin to those outlined in Equation (22). In line with recommendations from Wigger et al. [41], a weighting factor of $\theta = 0.55$ is used consistently across all pressure transient simulations.

A finite difference discretisation is applied to the partial derivatives present in Equations (18) and (20). The discretisation process yields a set of coupled nonlinear algebraic equations. These equations are formulated as follows:

$$\begin{aligned} G_1\left(p_m^{n+1}, p_{m+1}^{n+1}, v_m^{n+1}, v_{m+1}^{n+1}, \theta\right) = {} & \tfrac{1}{2\Delta t}\left[p_{m+1}^{n+1} + p_m^{n+1} - p_{m+1}^n - p_m^n\right] \\ & + \tfrac{\overline{v}}{\Delta x}\left[\theta\left(p_{m+1}^{n+1} - p_m^{n+1}\right) + (1-\theta)\left(p_{m+1}^n - p_m^n\right)\right] \\ & + \tfrac{\overline{\vartheta}_1\,\overline{\vartheta}_2}{\Delta x}\left[\theta\left(v_{m+1}^{n+1} - v_m^{n+1}\right) + (1-\theta)\left(v_{m+1}^n - v_m^n\right)\right] \end{aligned} \tag{23}$$

$$G_3\left(p_m^{n+1}, p_{m+1}^{n+1}, v_m^{n+1}, v_{m+1}^{n+1}, \theta\right) = \frac{1}{2\Delta t}\left[v_{m+1}^{n+1} + v_m^{n+1} - v_{m+1}^n - v_m^n\right] + \frac{\overline{v}}{\Delta x}\left[\theta\left(v_{m+1}^{n+1} - v_m^{n+1}\right) + (1-\theta)\left(v_{m+1}^n - v_m^n\right)\right] + \frac{\overline{\vartheta}_3}{\Delta x}\left[\theta\left(p_{m+1}^{n+1} - p_m^{n+1}\right) + (1-\theta)\left(p_{m+1}^n - p_m^n\right)\right] \quad (24)$$

A crucial aspect of this investigation is the air bubbles (void fraction, α) entrained by the rapid flow regime transitions at the downstream edge of the gas cavity in a pipe with a descending slope. In the numerical model, α—defined as the ratio of gas phase area to total cross-sectional area—is computed using Equation (25). It is essential to note that for this particular study, α values are treated as invariant. This assumption is based on negligible gas absorption or release during transient events.

$$\alpha = \frac{\beta}{1+\beta} \quad (25)$$

β is the air entrainment ratio for hydraulic jumps:

$$\beta = \frac{Q_a}{Q_w} \quad (26)$$

where Q_a is the volumetric airflow rate and Q_w is the volumetric water flow rate.

The air entrainment ratio for hydraulic jumps, denoted as β, was determined using Equation (27). Campbell and Guyton [61] initially proposed this relationship in 1953 to investigate air entrainment requirements arising from turbulent flow phenomena specifically the formation of hydraulic jumps within pressurised tunnels of dams located immediately downstream of gates. Subsequently, Safavi et al. [62] demonstrated that their experimental results for β in a gated tunnel with a circular geometry exhibited high concordance with the correlation developed by Campbell and Guyton [61]. In 1965, Wisner [63] independently arrived at the same relationship based on his own findings and recommended its use for computing air entrainment by hydraulic jumps. Further validation came in 1984 when Ahmed et al. [64] conducted extensive experimental studies in closed conduits with slopes ranging from horizontal to vertical, and found that his proposed equation coincided precisely with the equations put forth by the aforementioned researchers, confirming its applicability for hydraulic jump entrainment. More recently, Pozos et al. [65] and Pozos-Estrada [40] employed Equation (27) to evaluate air entrainment by hydraulic jumps in a real pumping pipeline and in an experimental apparatus with a circular cross-section, respectively. Both studies reported strong agreement between their experimental and numerical results. Based on the consistent validation and widespread applicability of this equation across various studies, the author of the present investigation elected to use this equation to compute the hydraulic jump entrainment coefficient, β.

$$\beta = 0.04(F_1 - 1)^{0.85} \quad (27)$$

where F_1 is the Froude number at the beginning of the hydraulic jump.

During simulations of fluid transients, a constant time step Δt is employed to maintain numerical stability. The spatial discretisation Δx is adaptively determined for each pipeline segment by the numerical model, based on the celerity of the water–air mixture a_{mix} (m/s). For a homogeneous water–air flow where the relative velocity between phases is negligible, the wave speed can be calculated using an expression similar to that proposed by Martin et al. [66]:

$$a_{mix} = \left[\frac{1}{\rho(1-\alpha)\left[\left(\frac{D\mu}{Ee}\right) + \left(\frac{\alpha}{E_a}\right) + \left(\frac{(1-\alpha)}{E_w}\right)\right]}\right]^{1/2} \quad (28)$$

In this context, E_a denotes the compressibility modulus of air (GPa), and μ indicates a constraint factor related to how much stress is transferred between fluid pressure and pipe deformation.

3. Experimental Apparatus and Data Acquisition Equipment

A setup was implemented to generate robust and reliable data, with the aim of validating the developed model. The setup comprises an upstream constant-head reservoir, which is a 2.3 m^3 hydropneumatic vertical tank. The test section, which simulates an undulating pipeline featuring a high point, is fabricated from transparent PVC piping with a diameter of 200 mm. This section is configured with two inclined pipe segments: an ascending pipe measuring 4.5 metres in length and a descending pipe of 7.5 metres in length. These segments are interconnected by a 60 cm long flexible plastic hose, also with a 200 mm internal diameter. The downstream end of the test section is connected to a gooseneck pipe that recirculates the flow back into an open tank. The choice of a 200 mm diameter was made to minimise the influence of surface tension effects, in accordance with recommendations provided by Pothof and Clemens [67,68] and Pothof [69]. The experimental apparatus is illustrated in Figure 3.

The filling of the test section is facilitated by a water tap connected to a 2.5 cm internal diameter hose, which is connected to a quarter-turn ball valve located at the upstream end of the ascending pipe. Additionally, smaller ball valves with an inner diameter of 1.25 cm are connected at regular intervals (every 10 cm) along the top of the test section. These valves enable air intake and exhaust during both filling and draining processes and serve as designated measurement points (tapping points). Furthermore, these tapping points can be used to inject compressed air via quick-connect fittings.

During the experimental investigations, two distinct methodologies were utilised to determine the water depths beneath the air pockets: (1) A direct measurement approach was employed using an acoustic depth gauge, which was inserted through the tapping points. Upon interacting with the water–air boundary, the device generated discernible acoustic output, and the sensor position was recorded to determine the depth. (2) An indirect measurement technique was also applied, involving a circumferential measurement method. The obtained measurements were subsequently adjusted to incorporate the pipe wall thickness, ensuring accuracy in experimental outcomes.

An ultrasonic flowmeter was utilised to quantify the water discharge within the experimental setup. To quantify frictional and singular pressure losses in the transparent PVC pipe sections, an open-ended water manometer was employed. The manometer was connected to pressure taps situated at the beginning and end of the test section using flexible plastic tubing. This configuration enabled pressure measurements to be taken both with and without the presence of air pockets. The differential pressure (ΔP) was calculated by measuring the vertical displacement (Δh) between the liquid menisci in each limb of the manometer.

A butterfly valve, controlled by a pneumatic actuator positioned at the outlet of the apparatus, was employed to generate transient flow conditions. Pressure transient measurements within the test section were obtained using two piezoelectric pressure transducers (PCB Model 113B26), each with a measurement range of 0 to 690 kPa and an error margin of $\pm$0.5%. These transducers were mounted at distances of 4.0 m and 5.4 m from the constant-head tank. The signals from the transducers were conditioned using a demodulator (Endevco Model 136) and then digitised via a data acquisition board (Data Translation DT9837 24-bit resolution and sampling rates up to 102.4 kS/s). The recorded pressure transient data from both transducers were put in storage on the disk drive of a computer for subsequent study. A pre-triggering routine was employed to capture the initial static pressure within the air pocket or gas cavity before the double-acting pneumatic actuator was activated, ensuring comprehensive recording of the transient pressure data.

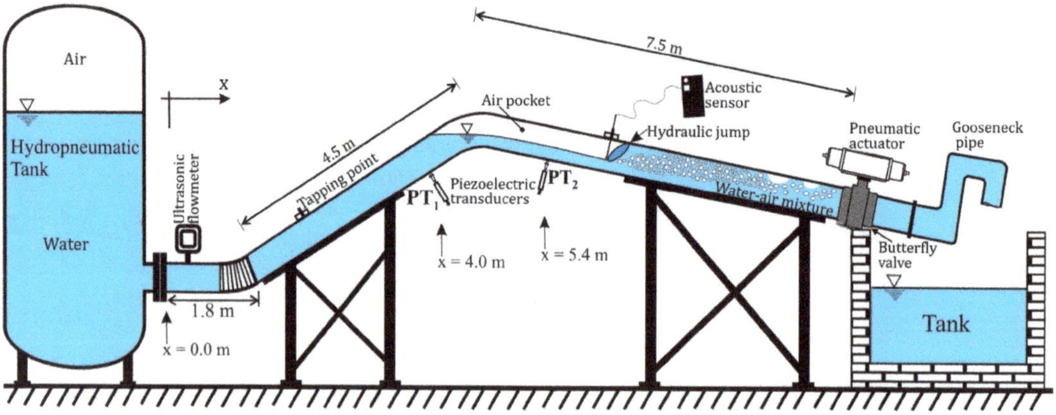

(a)

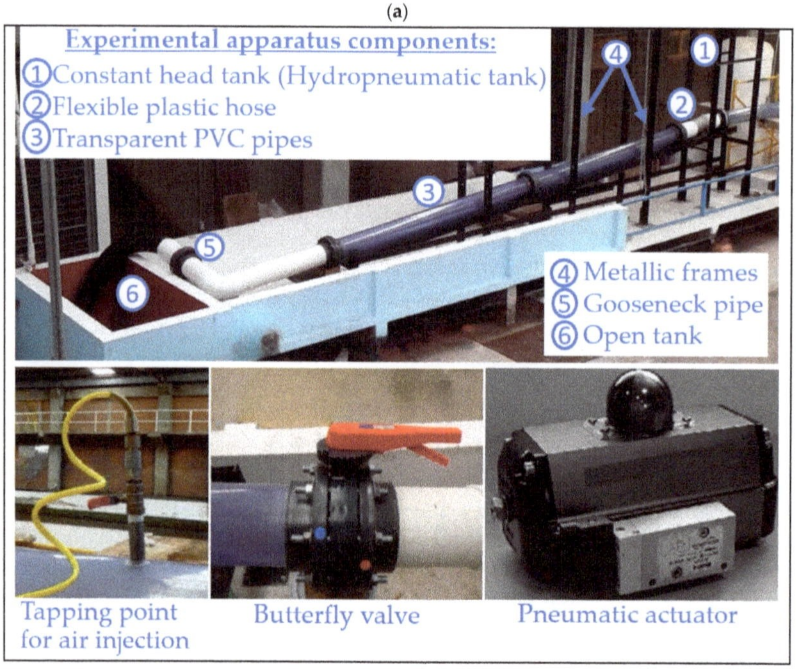

(b)

Figure 3. Setup. (**a**) Illustration of the apparatus. (**b**) Photograph of the experimental apparatus.

4. Experimental Procedure

An investigation into the response of gas cavities and two-component mixtures on pressure surges in pipelines was conducted. Four distinct experimental protocols were implemented using the testing apparatus. The datasets collected from these experiments were subsequently utilised in numerical simulations to model hydraulic transient scenarios both with and without entrapped air. Each experimental run within these protocols was replicated at least three times to minimise errors associated with experimentation.

Consistent results showed that the observed pressure transient oscillations exhibited regular periodicity and yielded reproducible pressure waveforms across all tests.

Throughout the experiments, a consistent hydraulic head of 139.7 kPa (20 psi) was sustained at the hydropneumatic vertical tank. Four distinct volumetric flow rates of water—0.020, 0.025, 0.030, and 0.035 m^3/s—and three different initial volumes of air pockets were tested. Additionally, to ensure optimal reservoir performance, the tank was kept with minimum air and water volumes of 0.3 m^3 and 2.0 m^3, respectively. Moreover, during the brief transient period lasting only a few seconds, there was no significant variation observed in the tank pressure.

Throughout the experiments, a consistent hydraulic head of 139.7 kPa (20 psi) was sustained at the upstream reservoir.

The experimental setup featured a test section with a compound slope, simulating a high point in an undulating pipeline profile. The upstream segment exhibited an adverse slope of $S_{up} = -0.532$, corresponding to an angle of $\phi = -28°$. In contrast, the downstream segment had a positive slope of $S_{down} = 0.466$, with an angle of $\phi = 25°$. During the experiments, stable air pockets consistently formed at the high point of the apparatus. These air pockets or gas cavities invariably ended in a rapid transition from super- to sub-critical flow (hydraulic jump) at the descending pipe section. This turbulence generated by the hydraulic jump entrains gas bubbles into the liquid phase, producing a water–air mixture characterised by discrete air bubbles dispersed throughout a continuous liquid medium (see Figure 1).

Extensive research has established that the presence of small gas cavities is capable of substantially increasing hydraulic transients, often to magnitudes that surpass the burst pressure limits of piping systems. Assuming that PVC possesses relatively low tensile strength and limited resistance to fatigue, especially under cyclic loading caused by transient pressures, a precautionary measure was taken. In each experimental run, an adequate amount of air was injected into the test section to deliberately form large and stable air pockets. This method was employed to reduce dynamic pressure fluctuations and thereby prevent catastrophic failure of the PVC piping.

Two distinct configurations for valve placement were examined. In experiments 1 and 2, the butterfly valve was situated at the outlet of the apparatus. In contrast, for experiments 3 and 4, it was positioned at the inlet. The four experimental protocols devised for this study are outlined in detail below.

Experiment 1: Pressurised flow conditions in the test section (no air accumulation).

In this experiment, the transparent pipe section was divided into two distinct segments with opposing slopes. The upstream segment was set at an upward slope of $S_{up} = -0.532$, while the downstream segment was configured with a downward slope of $S_{down} = 0.466$. The valve, integrated with a pneumatic actuator, was mounted at the downstream end of the apparatus. The procedure began by introducing water into the measurement section through a 2.5 cm diameter hose. Throughout the filling process, the absolute pressure in the upstream hydropneumatic tank was carefully maintained at a constant value of 139.7 kPa (20 psi). To ensure the complete removal of air, small 1.25 cm diameter quarter-turn valves, positioned at the top of the test section, were left open during the filling phase. Once the measurement section was entirely purged of air, the pressure in the hydropneumatic tank was confirmed to be stable, and the fluid discharge rates of water flowing across the pipe were recorded, the actuator was engaged to close the valve over a duration of 0.15 s, adhering to a linear valve closure.

Experiment 2: Stationary air pocket at the high point of the apparatus.

This experiment examined the impact of entrapped air on hydraulic transients produced by the rapid closure of a valve (0.15 s, linear closure). The methodology was as follows: With the test section operating under full flow conditions, air was introduced through a ball valve placed at the inlet of the test section. This resulted in the formation of a stationary gas cavity at the high point of the slope transition. A gooseneck pipe, installed beyond the valve, induced a hydraulic jump within the supercritical flow of the downward-sloping pipe. This created a bubbly flow regime downstream, where gas bubbles merged into larger bubbles that returned in counterflow without interruption. As a result, the air volume within the system remained constant throughout the test. Once the gas cavity stabilised at the high point, the water depths along the water–air boundary beneath the gas cavity were measured. The pressure in the hydropneumatic tank was then verified, and the water flow rate was recorded. Finally, the butterfly valve was closed to generate a hydraulic transient, which was captured and recorded using a computer system for subsequent processing and analysis.

The methodology employed for experiments 3 and 4 was similar to that used in the preceding two experiments, with the notable exception that the butterfly valve was positioned at the upstream end of the test section.

In experiments 2 and 4, the water depths at the point of hydraulic jump initiation were recorded to estimate both the entrained air content β and void fraction α. Furthermore, measurements were taken of air pocket lengths in the test section. Following each experimental run, the test section was thoroughly purged of air, after which the procedure was repeated with adjustments made to both water discharge rates and injected air volumes.

Table 1 provides a summary of the variables measured during experiments 2 and 4. These include air pocket volumes, air void fractions α, liquid depths at the beginning of the jumps, and the entrained air content β. Additionally, measurements were taken for volumetric flow rates, total pressure within the gas cavity under steady-state conditions, and the length of flow profiles beneath the air pockets.

Table 1. Variables measured during experiments 2 and 4.

Volume of Air	Liquid Depth	β	α	Total Pressure in the Air Pocket	Length of the Air Cavity Profiles	Length of the Air Cavity Profiles
(m^3)	y(m)	(-)	(%)	(kPa)	(m)	(m)
					Profile Upstream	Profile Downstream
			$Q_W = 0.02$ (m^3/s)			
0.0328	0.044	0.101	9.2	138.9	0.559	0.497
0.0381	0.041	0.117	10.4	139.7	0.559	0.675
0.0462	0.038	0.135	11.9	140.2	0.559	0.923
			$Q_W = 0.025$ (m^3/s)			
0.0323	0.046	0.116	10.4	138.8	0.452	0.581
0.0373	0.041	0.143	12.5	139.2	0.452	0.785
0.0471	0.038	0.164	14.1	139.8	0.452	1.101
			$Q_W = 0.03$ (m^3/s)			
0.0357	0.051	0.119	10.7	139.1	0.287	0.887
0.0473	0.042	0.165	14.2	139.6	0.287	1.191
0.0625	0.038	0.196	16.4	140.2	0.287	1.688

Table 1. *Cont.*

Volume of Air	Liquid Depth	β	α	Total Pressure in the Air Pocket	Length of the Air Cavity Profiles	Length of the Air Cavity Profiles
(m^3)	y(m)	(-)	(%)	(kPa)	(m)	(m)
					Profile Upstream	Profile Downstream
			$Q_W = 0.035$ (m^3/s)			
0.0279	0.060	0.104	9.4	139.9	0.203	0.721
0.0373	0.052	0.134	11.8	140.1	0.203	0.974
0.0478	0.050	0.145	12.7	140.5	0.203	1.331

5. Results

5.1. Results with the Valve at the Outlet of the Test Section

The discussion presents a comparative analysis of the experimental and numerical results derived from pressure transient analysis. Figure 4 illustrates the contrast of outcomes obtained from experimental measurements and simulations using the MOC following the closure of a butterfly valve. This particular simulation was conducted with a water-filled test section, intentionally excluding air entrapment at the high point, thereby validating the sole application of the MOC. A slight overestimation of pressure amplitudes is noted in the results achieved with the model in comparison with the experimental outcomes. This discrepancy may be attributed to measurement uncertainties inherent in the experimental setup. Nonetheless, despite this minor deviation, a robust correlation between the numerical and experimental findings is evident, demonstrating a high level of agreement overall.

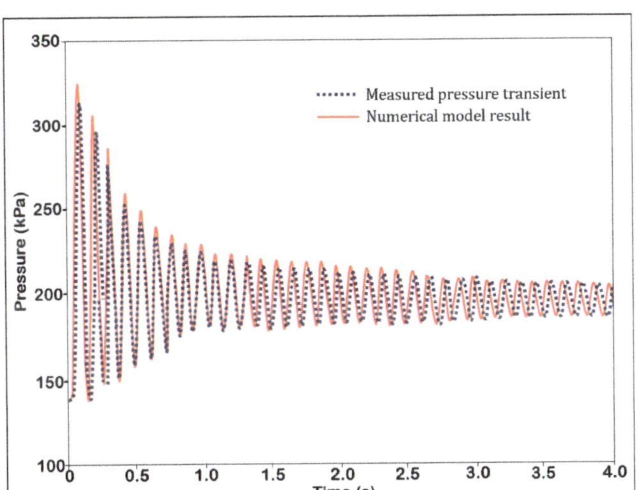

Figure 4. Comparison of measured and computed results without air, with the valve at the outlet of the test section.

The influence of varying volumes of gas cavities and the downstream water–air mixture void fraction on pressure transients during test runs, conducted with a constant liquid flow rate of 0.02 m^3/s, is depicted in Figure 5. The analysis indicates that the increase in gas cavity volumes and void fraction considerably reduces the magnitude of pressure surges while simultaneously extending the wave period. Experimental validation using piezoelectric transducers produced consistent results, supporting the observed trends. Moreover, numerical simulations of hydraulic transients—both with and without entrapped air pockets—across various water flow discharges exhibited similar pressure wave patterns. As a result, only a representative selection of the simulation outcomes is presented in this study.

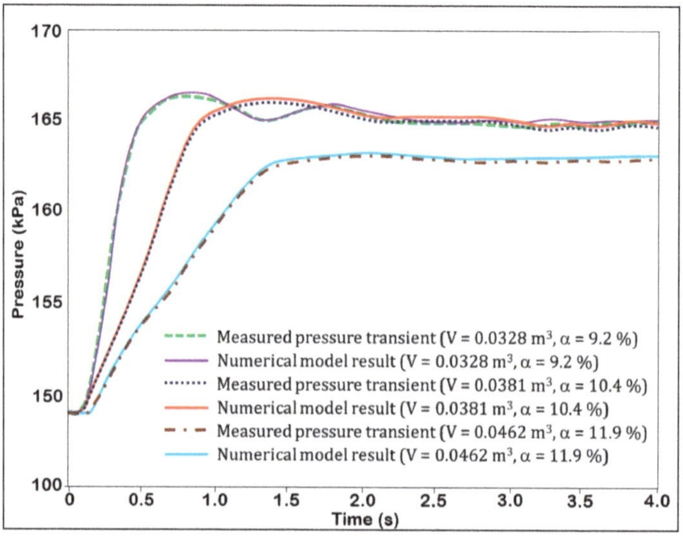

Figure 5. Comparison of measured and computed results varying volumes of air pockets and void fractions, with the valve at the outlet of the test section.

Figure 6 provides a comparative analysis of the simulated fluid transient behaviour within a test section that contains a gas cavity volume of $V = 0.0328$ m^3, with a downstream water–air mixture of 9.2%. The outcomes obtained with the model are contrasted with experimental pressure transient measurements. The analysis reveals a significant reduction in the magnitude of pressure variations owing to the presence of the gas cavity. Additionally, the downstream water–air mixture contributes to a further substantial decrease in the pressure transient envelope compared to conditions where no air is present.

The analysis of the data shown in Figures 7 and 8 indicates a robust correlation between the water–air mixtures and the behaviour of pressure transients. In particular, an increase in both the air in the mixtures and the volume of gas pockets is associated with a notable decrease in the magnitude of pressure oscillations throughout the test section. Figure 8 presents an examination of the numerical simulations, highlighting a significant reduction in the reflection of transient pressure waves when an air-water mixture is present. In this instance, an air volume of 0.0462 m^3, along with its corresponding air void fraction of 10.4%, demonstrates a beneficial damping effect.

The analysis of the preceding graphical data suggests that the presence of gas cavities, along with a downstream void fraction, markedly reduces pressure transients within the test section. Additionally, the observed damping of pressure fluctuations can be attributed to the considerable air volumes and associated void fractions, leading to a more pronounced attenuation of transients compared to scenarios without any air content.

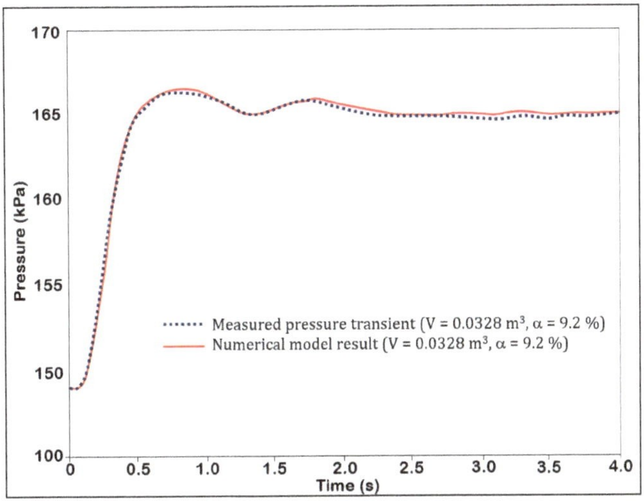

Figure 6. Comparison of measured and computed results with air, with the valve at the outlet of the test section. Steady passage occurs at approximately 40 s.

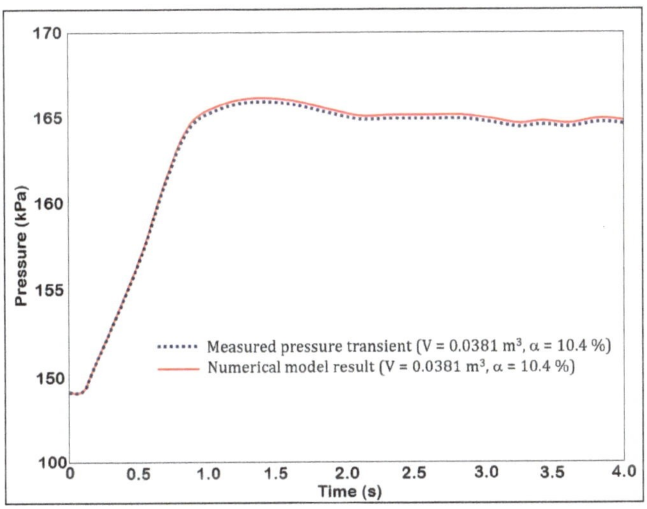

Figure 7. Comparison of measured and computed results with air, with the valve at the outlet of the test section. Steady passage occurs at approximately 40 s.

The observed attenuation of both experimental and simulated pressure transients can be associated with the effect of air pocket volume on the characteristics of hydraulic jumps. An increase in the volume of the air pocket results in a decrease in the water depth at the point of hydraulic jump initiation (as shown in Table 1). This reduction in depth enhances turbulence within the jump, promoting the entrainment of gas bubbles from the gas cavity. The presence of this air-water mixture significantly decreases wave celerity, thereby diminishing the propagating pressure wave. This phenomenon has been substantiated by prior research findings [38,70,71].

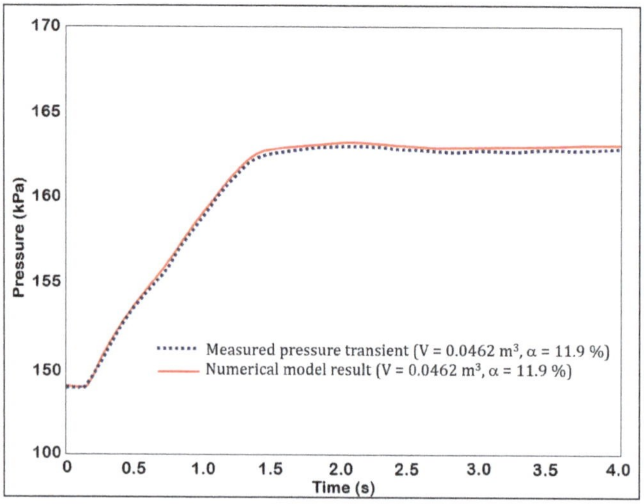

Figure 8. Comparison of measured and computed results with air, with the valve at the outlet of the test section. Steady passage occurs at approximately 40 s.

5.2. Results with the Valve at the Inlet of the Test Section

Consistent with the approach taken for hydraulic transient simulations involving the valve positioned at the downstream end, only a representative selection of simulations for the valve located at the upstream end is presented here. These simulations focus on a water flow rate of 0.02 m^3/s and explore variations in both the volume of trapped gas pockets and the air void fraction.

Figure 9 presents a comparative analysis of the measured and numerically simulated pressure transients within the test section when the valve closes suddenly at the upstream end. The numerical model, which utilises the MOC, was run without considering trapped gas pockets at the test section. Both the experimental and simulated data exhibited an immediate drop in pressure from the steady-state condition upon valve closure. Although the computed pressure amplitudes were marginally greater in comparison with the experimental outcomes, the discrepancy can be attributed to the inherent uncertainties associated with experimental measurements. Nonetheless, a satisfactory correlation between the numerical and experimental results is evident.

Figure 10 shows that numerical and experimental results obtained with the valve positioned at the upstream edge of the experimental apparatus demonstrate the initial pressure peak is considerably greater in the presence of air. Specifically, with a gas cavity volume of 0.328 m^3 and an air void fraction of 9.2%, the initial pressure peak reached an initial pressure peak of 189.5 kPa. This represents a substantial increase compared to the 76.3 kPa observed under conditions without trapped air. Furthermore, it was noted that increasing the volume of air pockets and the air void fraction contributes to a reduction in the magnitude of pressure oscillations. In this particular case, the presence of air pockets provides a degree of protection against severe negative pressure events. It can be concluded that air pockets function effectively as air chambers, eliminating the risk of cavitation and preventing potential pipe collapse.

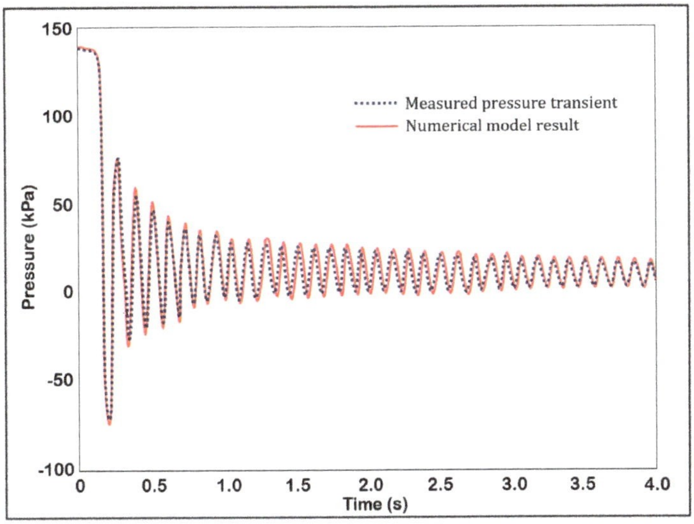

Figure 9. Comparison of measured and computed results without air, with the valve at the inlet of the test section.

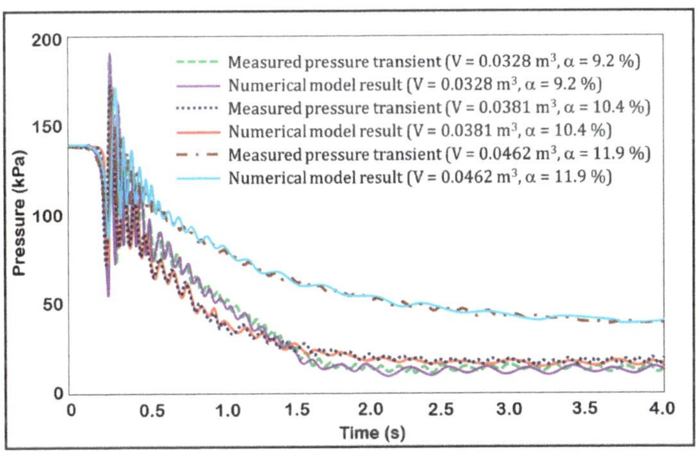

Figure 10. Comparison of measured and computed results varying volumes of air pockets and air void fractions, valve at the inlet of the test section.

Figures 11–13 demonstrate that an increase in both the gas cavity volume and the void fraction results in a reduction in the amplitude of pressure transients. Nevertheless, the initial peak pressure value was observed to be greater in scenarios involving air pockets compared to those without air. This indicates that the presence of gas cavities and the consequent void fraction partially attenuate the propagation of the pressure wave, thereby influencing the dynamic response of the system.

An examination of the computational results presented in Figure 11 reveals that a gas cavity of 0.328 m^3, corresponding to an air void fraction of 9.2%, generates the highest initial peak pressure of 189.5 kPa. This value represents a substantial increase from the 76.3 kPa recorded without an entrapped gas cavity (i.e., full pipe flow). The amplification of this pressure peak is primarily attributed to the dynamic compression and expansion of the gas cavity, combined with reflections of the transient pressure wave at the butterfly valve, the gas cavity interface, and the downstream edge of the test section. These interactions collectively contribute to the observed pressure surge.

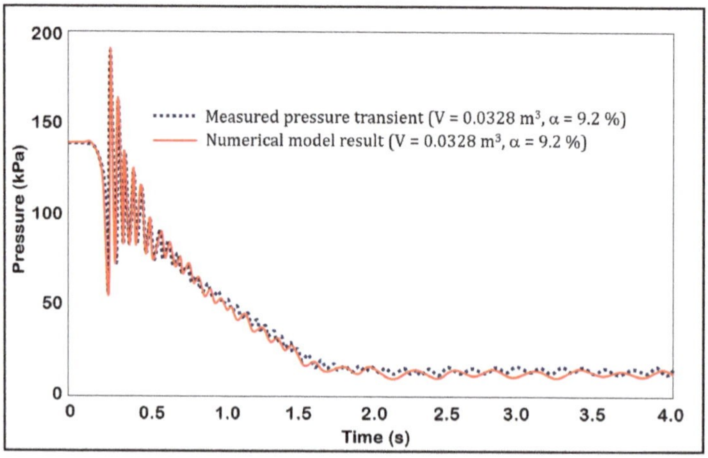

Figure 11. Comparison of measured and computed results with air, with the valve at the inlet of the test section. Steady passage occurs at approximately 55 s.

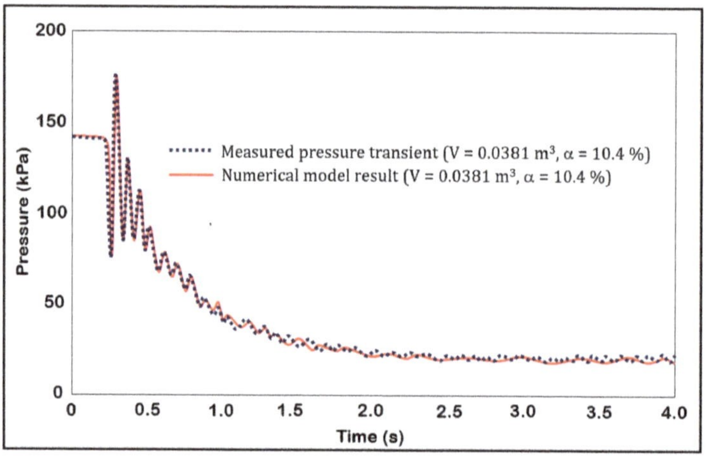

Figure 12. Comparison of measured and computed results with air, with the valve at the inlet of the test section. Steady passage occurs at approximately 55 s.

Figures 11–13 demonstrate a notable increase in the initial pressure peak. However, it is worth mentioning that the existence of gas pockets and their corresponding air void fractions play a significant role in alleviating pressure drops. In scenarios devoid of air, a minimum pressure of −74.7 kPa was recorded, while when air was present, gas pockets worked as air chambers, elevating the lowest pressure to less negative values when compared to conditions without air.

The analysis of transient pressures highlights the necessity of safeguarding the pipeline against both overpressure and negative pressure events. To address these risks effectively, the strategic installation of air/vacuum valves is strongly advised. These valves should be positioned at the inlet of the piping network, preceding the flow control valve, as well as at other key locations such as high points along the pipeline. A typical example of their application is at the discharge point of a pump. The air/vacuum valves facilitate the release or admission of air as required, thereby preventing issues arising from air pockets within the pipeline.

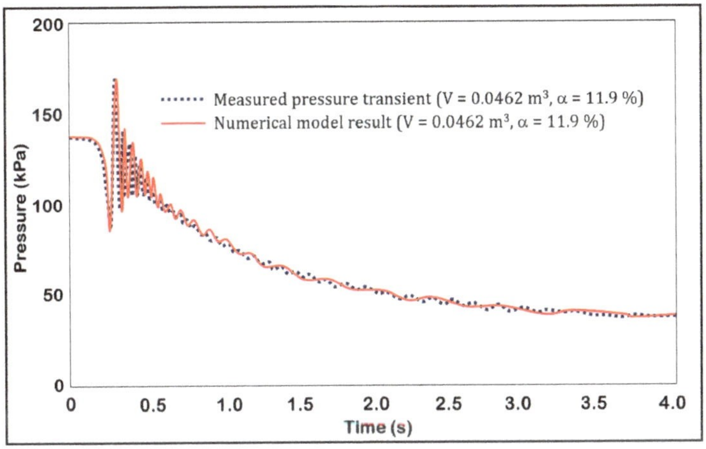

Figure 13. Comparison of measured and computed results with air, with the valve at the inlet of the test section. Steady passage occurs at approximately 55 s.

Experimental and numerical results have revealed that the positioning of the valve plays a pivotal role in influencing transient pressure dynamics within pipeline systems. The findings highlight that pressure waves undergo significant damping as they propagate through the turbulent zone downstream of the gas pocket in the bubbly flow regime. This turbulence, particularly at the hydraulic jump, markedly reduces the wave velocity and attenuates the amplitude of pressure peaks. Importantly, an increase in air volume intensifies turbulence and agitation at the hydraulic jump, resulting in the detachment of additional bubbles from the primary air pocket. This heightened turbulence further decelerates the wave speed and diminishes the magnitude of pressure peaks. As a result, when the valve is situated downstream, pressure waves are effectively attenuated. In contrast, positioning the valve upstream leads to an increase in pressure peak magnitudes.

6. Conclusions

In pipeline systems with undulating topography, air pockets frequently accumulate at elevated sections. Hydraulic jumps typically form at the downstream end of these gas cavities, generating turbulent energy at the jump interface. The turbulence entrains small air bubbles into the flowing liquid, creating a two-phase bubbly flow regime characterised by a dispersed air phase within a continuous liquid phase. This research introduces a numerical model that accounts for the integrated influence of gas cavities and the resulting void fractions on pressure surge behaviour within pipelines. The validity of the numerical model is confirmed through comparisons with experimental data collected from a specially designed hydraulic transient test rig.

The impact of air pocket volume and subsequent void fraction on pressure surges was investigated through both numerical simulations and experimental techniques. Transient conditions were initiated when a valve was closed suddenly. Experiments were conducted with the valve situated both at the outlet and inlet of the test section, with the air pocket volume systematically varied for each valve position.

The outcomes obtained from laboratory testing and computational modelling with the valve positioned downstream demonstrate a clear attenuation of pressure surges owing to the existence and characteristics of gas cavities. Specifically, gas cavities consistently decrease the magnitude of hydraulic transients. The magnitude of this damping effect is directly correlated with the volume of the gas cavity and the resulting water–air mixture. As shown in Figures 6–8, increasing both air volume and air void fraction leads to a more pronounced reduction in pressure transients. This attenuation is attributed to the rapid transition from super- to sub-critical flow at the descending pipe section, which is promoted by the gas cavity. The enhanced turbulence leads to increased entrainment of gas bubbles, forming a water–air mixture that significantly reduces the wave celerity, thereby dissipating the energy of the propagating pressure wave. These findings confirm that strategically introducing and managing air pockets upstream of a control valve can be an effective strategy for mitigating transient pressure surges in pipeline systems.

The results obtained with the valve positioned upstream reveal a more complex interaction between air pockets and transient pressures. While the presence of air pockets generally reduces the overall amplitude of pressure oscillations and mitigates negative pressure events, the initial pressure peak can be significantly amplified compared to scenarios without air. As illustrated in Figures 11–13, increasing the volume of the gas cavity and the percentage of gas in the water–air mixture contributes to a reduction in the magnitude of pressure oscillations. However, the first pressure peak is consistently higher in the presence of gas, with a gas cavity volume of 0.328 m^3 and corresponding air void fraction of 9.2% inducing the highest initial peak pressure of 189.5 kPa, a substantial increase from the 76.3 kPa observed in full pipe flow (i.e., without air entrapment). This amplification is attributed to the dynamic compression and expansion of the gas cavity, coupled with reflections of the pressure wave at the valve, air pocket interface, and downstream boundary (Gooseneck pipe). Crucially, however, the presence of pockets provides a degree of protection against severe negative pressure events. The minimum pressure recorded in the absence of air (−74.7 kPa) is substantially lower than the minimum pressures observed when air pockets are present, thereby indicating that these air pockets effectively function as air chambers, mitigating the potential for cavitation and pipeline collapse.

The findings emphasise the importance of protecting pipelines against both overpressure and sub-atmospheric pressure conditions. The strategic positioning of air/vacuum valves at critical locations is strongly recommended. These locations should include positions upstream of flow control valves and at high points along the pipeline. Such valves facilitate the controlled release or ingress of air as required, thus mitigating problems arising from air entrapment and ensuring the operational stability of the pipeline system. This preventative measure is particularly crucial in systems where transient pressure events are anticipated.

Funding: This research received no external funding.

Data Availability Statement: The original contributions presented in this study are included in the article. Further inquiries can be directed to the corresponding author.

Conflicts of Interest: The author declare no conflict of interest.

Notation

The following notations are used in this manuscript:

A	internal transverse area of the pipe (m^2)
a_{mix}	celerity of the water–air mixture (m/s)
c	velocity of wave transmission for water alone (m/s)
C	constant determined from equilibrium conditions of the gas cavity
D	diameter of the pipe (m)

e	wall thickness of the pipe (m)
E	elastic modulus of the material constituting the pipe (GPa)
E_a	compressibility modulus of air (GPa)
E_w	compressibility modulus of water (GPa)
f	Darcy–Weisbach resistance coefficient
F_1	Froude number at the beginning of the hydraulic jump
g	gravitational acceleration (m/s^2)
H	piezometric line (m)
H_A	total hydraulic energy (m)
H_b	barometric pressure head (m)
p	pressure at a specified cross-section (Pa)
Q_a	volumetric airflow rate
Q_w	volumetric water flow rate
t	time (s)
v	velocity of water within the pipe (m/s)
V	volume of the entrapped gas cavity (m^3)
v_{mix}	mixture velocity is denoted by (m/s)
v_g	velocity of the gas phase (m/s)
x	longitudinal position along the pipe (m)
z	vertical displacement from the centreline of the pipe to a reference datum (m)
α	air void fraction
β	air entrainment ratio for hydraulic jumps
ε	polytropic exponent
ϕ	angle at which the pipe is inclined, measured from the horizontal plane
μ	constraint factor
θ	weighting factor
ρ	density of the liquid (kg/m^3)
ρ_g	density of the gas (kg/m^3)
ρ_{mix}	density of the water–air mixture (kg/m^3)
τ_o	boundary shear stress (Pa)

Appendix A

The coefficients within the Equations (18) and (19) are presented as follows:

$$C_1 = \alpha(1-\alpha)\left(\frac{1}{E_a} - \frac{1}{E_w}\right)\left[\frac{D\mu}{Ee} + \frac{\alpha}{E_a} + \frac{(1-\alpha)}{E_w}\right]^{-1} \tag{A1}$$

$$C_2 = \left[\alpha(1-\alpha)\left(\frac{1}{E_a} - \frac{1}{E_w}\right)\right]^{-1} \tag{A2}$$

$$C_3 = \frac{1}{\rho(1-\alpha)} \tag{A3}$$

$$b = gsin\theta - \frac{f}{2D}v_{mix}|v_{mix}| \tag{A4}$$

References

1. Fox, J.A. *Hydraulic Analysis of Unsteady Flow in Pipe Networks*; MacMillan Press: London, UK, 1977.
2. Falvey, H.T. *Air-Water Flow in Hydraulic Systems, Engineering Monograph No. 41*; Bureau of Reclamation: Washington, DC, USA, 1980.
3. Qiu, D.Q. Transient Analysis and the Effect of Air Pockets in a Pipeline. Master's Thesis, University of Liverpool, Liverpool, UK, 1995.
4. Wisner, P.E.; Mohsen, F.N.; Kouwen, N. Removal of Air from Water Lines by Hydraulic Means. *J. Hyd. Div.* **1975**, *101*, 243–257.
5. Richards, R.T. Air Binding in Water Pipelines. *J. AWWA* **1962**, *54*, 719–730. [CrossRef]
6. Edmunds, R.C. Air Binding in Pipes. *J. AWWA* **1979**, *71*, 273–277. [CrossRef]

7. Sailer, R.E. San Diego Aqueduct. *J. Civil. Eng.* **1955**, *25*, 38–40.
8. Thomas, S. *Air Management in Water Distribution Systems: A New Understanding of Air Transfer*; Clear Water Legacy: Burlington, ON, Canada, 2003.
9. Sun, J.; Wang, R.; Duan, H.F. Multiple-fault detection in water pipelines using transient based time-frequency analysis. *J. Hydroinformatics* **2016**, *18*, 975–989.
10. Pozos, O.; Giesecke, J.; Marx, W.; Rodal, E.A.; Sanchez, A. Experimental Investigation of Air Pockets in Pumping Pipeline Systems. *J. Hydraul. Res.* **2010**, *48*, 269–273. [CrossRef]
11. Pozos, O. Investigation on the Effects of Entrained Air in Pipelines. Ph.D. Thesis, University of Stuttgart, Stuttgart, Germany, 2007.
12. Walski, T.M.; Barnhart, T.; Driscoll, J.; Yencha, R. Hydraulics of Corrosive Gas Pockets in Force Mains. *Water Environ. Res.* **1994**, *66*, 772–778. [CrossRef]
13. Zhou, F.; Hicks, F.E.; Steffler, P.M. Transient flow in a rapidly filling horizontal pipe containing trapped air. *J. Hydraul. Eng.* **2002**, *128*, 625–634. [CrossRef]
14. Carlos, M.; Arregui, F.G.; Cabrera, E.; Palau, V. Understanding Air Release Through Air Valves. *J. Hydraul. Eng.* **2011**, *137*, 461–469. [CrossRef]
15. Balacco, G.; Apollonio, C.; Piccinni, A.F. Experimental Analysis of Air Valve Behaviour During Hydraulic Transients. *J. Appl. Water Eng. Res.* **2015**, *3*, 3–11. [CrossRef]
16. Arregui, F.; Garcia, J.; Kruisbrink, A.; Cabrera, E.; Fuertes, V.S.; Palau, C.V.; Gascón, L. Air valves dynamic behaviour. In Proceedings of the PEDS 2003–Pumps, Electromechanical Devices and Systems Applied to Urban Water Management, Valencia, Spain, 22–25 April 2003.
17. Kruinsbrick, A.C.H.; Arregui, F.; Carlos, M.; Bergant, A. Dynamic performance characterization of air valves. In Proceedings of the 9th International Conference on Pressure Surges, Chester, UK, 24–26 March 2004; pp. 33–48.
18. Lee, N.H. Effect of Pressurization and Expulsion of Entrapped Air in Pipelines. Ph.D. Thesis, Georgia Institute of Technology, Atlanta, GA, USA, 2005.
19. Izquierdo, J.; Fuertes, V.S.; Cabrera, E.; Iglesias, P.L.; Garcia-Serra, J. Pipeline start-up with entrapped air. *J. Hydraul. Res.* **1999**, *37*, 579–590. [CrossRef]
20. Apollonio, C.; Balacco, G.; Fontana, N.; Giugni, M.; Marini, G.; Piccinni, A.F. Hydraulic Transients Caused by Air Expulsion During Rapid Filling of Undulating Pipelines. *Water* **2016**, *8*, 25. [CrossRef]
21. Aguirre-Mendoza, A.M.; Oyuela, S.; Espinoza-Román, H.G.; Coronado-Hernández, O.E.; Fuertes-Miquel, V.S.; Paternina-Verona, D.A. 2D CFD Modeling of Rapid Water Filling with Air Valves Using OpenFOAM. *Water* **2021**, *13*, 3104. [CrossRef]
22. Aguirre-Mendoza, A.M.; Paternina-Verona, D.A.; Oyuela, S.; Coronado-Hernández, O.E.; Besharat, M.; Fuertes-Miquel, V.S.; Iglesias-Rey, P.L.; Ramos, H.M. Effects of Orifice Sizes for Uncontrolled Filling Processes in Water Pipelines. *Water* **2022**, *14*, 888. [CrossRef]
23. Hurtado-Misal, A.D.; Hernández-Sanjuan, D.; Coronado-Hernández, O.E.; Espinoza-Román, H.; Fuertes-Miquel, V.S. Analysis of Sub-Atmospheric Pressures during Emptying of an Irregular Pipeline without an Air Valve Using a 2D CFD Model. *Water* **2021**, *13*, 2526. [CrossRef]
24. Fuertes-Miquel, V.S. Hydraulic Transients in Water Distribution Systems. *Water* **2022**, *14*, 3612. [CrossRef]
25. Ramos, H.M.; Fuertes-Miquel, V.S.; Tasca, E.; Coronado-Hernández, O.E.; Besharat, M.; Zhou, L.; Karney, B. Concerning dynamic effects in pipe systems with two-phase flows: Pressure surges, cavitation, and ventilation. *Water* **2022**, *14*, 2376. [CrossRef]
26. Tasca, E.; Karney, B.; Fuertes-Miquel, V.S.; Dalfré Filho, J.G.; Luvizotto, E., Jr. The Crucial Importance of Air Valve Characterization to the Transient Response of Pipeline Systems. *Water* **2022**, *14*, 2590. [CrossRef]
27. Martin, C.S. Entrapped Air in Pipelines. In Proceedings of the Second International Conference on Pressure Surges, London, UK, 22–24 September 1976; pp. 15–28.
28. Martin, C.S. Two-phase gas-liquid experiences in fluid transients. In Proceedings of the 7th International Conference on Pressure Surge and Fluid Transients in Pipelines and Open Channels, Harrogate, UK, 16–18 April 1996; pp. 65–81.
29. Lauchlan, C.S.; Escarameia, M.; May, R.W.P.; Borrows, R.; Gahan, C. *Air in Pipelines–A Literatur Review*; Report SR649; HR Wallingford: Wallingford, UK, 2005.
30. Jönsson, L. Maximum transient pressures in a conduit with check valve and air entrainment. In *Proceeding of the International Conference on the Hydraulics of Pumping Stations, Manchester, UK, 17–19 September 1985*; British Hydromechanics Research Association: Cranfield, UK, 1985; pp. 55–76.
31. Jönsson, L. Anomalous pressure transients in sewage lines. In Proceedings of the International Conference on Unsteady Flow and Transients, Durham, UK, 29 September–1 October 1992; pp. 251–258.
32. Fuertes-Miquel, V.S.; Coronado-Hernández, O.E.; Mora-Meliá, D.; Iglesias-Rey, P.L. Hydraulic modeling during filling and emptying processes in pressurized pipelines: A literature review. *Urban Water J.* **2019**, *16*, 299–311.

33. Burrows, R. A cautionary note on the operation of pumping mains without appropriate surge control and the potentially detrimental impact of small air pockets. In Proceedings of the IWA International Conference, Valencia, Spain, 22–25 April 2003; pp. 22–25.
34. Qiu, D.Q.; Borrows, R. Prediction of pressure transients with entrapped air in a pipeline. In Proceedings of the 7th International Conference on Pressure Surge and Fluid Transients in Pipelines and Open Channels, BHRA, Harrogate, UK, 16–18 April 1996; pp. 251–263.
35. Stephenson, D. Effects of air valves and pipework on water hammer pressure. *J. Transp. Eng.* **1997**, *123*, 101–106.
36. Thorley, A.R.D. *Fluid Transients in Pipeline Systems*, 2nd ed.; D. & L. George Ltd.: London, UK, 2004.
37. Gahan, C.M. A Review of the Problem of Air Release/Collection in Water Pipelines with In-Depth Study of the Effects of Entrapped Air on Pressure Transients. Master's Thesis, University of Liverpool, Liverpool, UK, 2004.
38. Pearsall, I.S. The velocity of water hammer waves, Symposium on Surge in Pipelines. *Inst. Mech. Eng.* **1965**, *180*, 12–20.
39. Ngoh, K.L.; Lee, T.S. Air influence on pressure transients with air vessel. In Proceedings of the XIX Symposium on Hydraulic Machinery and Cavitation, IAHR, Singapore, 9–11 September 1998; pp. 665–672.
40. Pozos-Estrada, O. Investigation of the combined effect of air pockets and air bubbles on fluid transients. *J. Hydroinformatics* **2018**, *20*, 376–392. [CrossRef]
41. Wiggert, D.C.; Martin, C.S.; Naghash, M.; Rao, P.V. Modeling of transient two-component flow using a four-point implicit method. In *Numerical Methods for Fluid Transient Analysis; Proceedings of the Applied Mechanics, Bioengineering, and Fluids Engineering conference, Houston, TX, 20–22 June 1983*; American Society of Mechanical Engineers: New York, NY, USA, 1983; pp. 23–28.
42. Amein, M.; Fang, C.S. Implicit flood routing in natural channels. *J. Hydraul. Div.* **1970**, *96*, 2481–2500.
43. Amein, M.; Chu, H.L. Implicit numerical modeling of unsteady flows. *J. Hydraul. Div.* **1975**, *101*, 717–731. [CrossRef]
44. Almeida, A.B.; Ramos, H.M. Water supply operation: Diagnosis and reliability analysis in a Lisbon pumping system. *J. Water Supply Res. Technol.—AQUA* **2010**, *59*, 66–78.
45. Burrows, R.; Qiu, D.Q. Effect of air pockets on pipeline surge pressure. In *Proceedings of the Institution of Civil Engineers-Water Maritime and Energy*; Emerald Publishing Limited: Leeds, UK, 1995; Volume 112, pp. 349–361.
46. Ivetic, M. Forensic transient analyses of two pipeline failures. *Urban Water J.* **2004**, *1*, 85–95.
47. Soares, A.K.; Covas, D.I.; Reis, L.F.R. Leak detection by inverse transient analysis in an experimental PVC pipe system. *J. Hydroinformatics* **2011**, *13*, 153–166. [CrossRef]
48. Zhou, F.; Hicks, F.E.; Steffler, P.M. Observations of air–water interaction in a rapidly filling horizontal pipe. *J. Hydraul. Eng.* **2002**, *128*, 635–639. [CrossRef]
49. Vasconcelos, J.G.; Wright, S.J. Rapid flow startup in filled horizontal pipelines. *J. Hydraul. Eng.* **2008**, *134*, 984–992.
50. Martin, C.S.; Lee, N. Measurement and rigid column analysis of expulsion of entrapped air from a horizontal pipe with an exit orifice. In Proceedings of the 11th International Conference on Pressure Surges, Lisbon, Portugal, 24–26 October 2012; pp. 527–542.
51. Kaur, K.; Laanearu, J.; Annus, I. Air pocket dynamics under bridging of stratified flow during rapid filling of a horizontal pipe. *J. Hydraul. Eng.* **2023**, *149*, 04022030.
52. Chaudhry, M.H. *Applied Hydraulic Transients*, 2nd ed.; Van Nostrand Reinhold: New York, NY, USA, 1987.
53. Wylie, E.B.; Streeter, V.L.; Suo, L. *Fluid Transients in Systems*; Prentice Hall: Englewood Cliffs, NJ, USA, 1993.
54. Lee, N.H.; Martin, C.S. Experimental and Analytical Investigation of Entrapped Air in a Horizontal Pipe. In Proceedings of the 3rd ASME/JSME Joint Fluids Engineering Conference, San Francisco, CA, USA, 18–23 July 1999.
55. Martin, C.S.; Lee, N.H. Rapid Expulsion of Entrapped Air through an Orifice. In Proceedings of the 8th International Conference on Pressure Surges-Safe Design and Operation of Industrial Pipe Systems, The Hague, The Netherlands, 12–14 April 2000; pp. 125–132.
56. De Martino, G.; Giugni, M.; Viparelli, M.; Gisonni, C. Pressure Surges in Water Mains Caused by Air Release. In Proceedings of the 8th International Conference on Pressure Surges-Safe Design and Operation of Industrial Pipe Systems, The Hague, The Netherlands, 12–14 April 2000; pp. 147–159.
57. Fuertes, V.S.; Arregui, F.; Cabrera, E.; Iglesias, P.L. Experimental setup of entrapped air pockets model validation. In Proceedings of the 8th International Conference on Pressure Surges-Safe Design and Operation of Industrial Pipe Systems, The Hague, The Netherlands, 12–14 April 2000; pp. 133–146.
58. Chaudhry, M.H.; Bhallamudi, S.M.; Martin, C.S.; Naghash, M. Analysis of transient pressures in bubbly, homogeneous, gas-liquid mixtures. *J. Fluids Eng.* **1990**, *112*, 225–231.
59. Yadigaroglu, G.; Lahey, R.T., Jr. On the Various Forms of the Conservation Equations in Two-Phase Flow. *Int. J. Multiph. Flow* **1976**, *2*, 477–494.
60. Fread, D.L. *Numerical Properties of Implicit Four-Point Finite Difference Equations of Unsteady Flow (Vol. 18)*; US Department of Commerce, National Oceanic and Atmospheric Administration, National Weather Service, Office of Hydrology: Washington, DC, USA, 1974.

61. Campbell, F.B.; Guyton, B. Air demand in gated outlet works. In Proceedings of the Minnesota International Hydraulic Convention, Minneapolis, MN, USA, 1–4 September 1953; ASCE: Reston, VA, USA, 1953; pp. 529–533.
62. Safavi, K.; Zarrati, A.R.; Attari, J. Experimental study of air demand in high head gated tunnels. In Proceedings of the Institution of Civil Engineers-Water Management; Emerald Publishing Limited: Leeds, UK, 2008; Volume 161, pp. 105–111.
63. Wisner, P. On the role of the Froude criterion for the study of air entrainment in high velocity flows. In Proceedings of the 11th international association for hydraulic research (IAHR Congress), Madrid, Spain, 3–5 September 1965; Volume 611.
64. Ahmed, A.A.; Ervine, D.A.; McKeogh, E.J. The Process of Aeration in Closed Conduit Hydraulic Structures. In Proceedings of the a Symposium on Scale Effects in Modelling Hydraulic Structures, Esslingen am Neckar, Germany, 3–6 September 1984; Kobus, H., Ed.; Technische Akademie Esslingen: Ostfildern, Germany, 1984; pp. 1–11.
65. Pozos, O.; Sanchez, A.; Rodal, E.A.; Fairuzov, Y.V. Effects of water–air mixtures on hydraulic transients. *Can. J. Civ. Eng.* **2010**, *37*, 1189–1200.
66. Martin, C.S.; Padmanabhan, M.; Wiggert, D.C. Pressure Wave Propagation in Two-Phase Bubbly Air-Water Mixtures. In Proceedings of the 2nd International Conference on Pressure Surges, London, UK, 22–24 September 1976; pp. 1–16.
67. Pothof, I.W.M.; Clemens, F.H.L.R. On Elongated Air Pockets in Downward Sloping and Inclined Pipes. *J. Hydraul. Res.* **2010**, *48*, 499–503.
68. Pothof, I.W.M.; Clemens, F.H.L.R. Experimental Study of Air-Water Flow in Downward Sloping Pipes. *Int. J. Multiph. Flow* **2011**, *37*, 278–292.
69. Pothof, I.W.M. Co-current Air-Water Flow in Downward Sloping Pipes, Transport of Capacity Reducing Gas Pocket in Wastewater Mains. Ph.D. Thesis, Delft University of Technology, Delft, The Netherlands, 2011.
70. Davis, M.R. Structure and analysis of gas-liquid mixture flow. In Proceedings of the 7th Australasian Conference on Hydraulics and Fluid Mechanics, Brisbane, Australia, 18–22 August 1980; Institution of Engineers: Brisbane, Australia, 1980; pp. 416–419.
71. Ewing, D.J.F. Allowing for free air in waterhammer analysis. In Proceedings of the 3rd International Conference on Pressure Surge, BHRA, Canterbury, UK, 25–27 March 1980; pp. 127–146.

water

MDPI

Article

Application of Newton–Raphson Method for Computing the Final Air–Water Interface Location in a Pipe Water Filling

Dalia M. Bonilla-Correa [1], Óscar E. Coronado-Hernández [2,*], Vicente S. Fuertes-Miquel [3], Mohsen Besharat [4] and Helena M. Ramos [5]

1 Facultad de Ciencias Exactas y Naturales, Universidad de Cartagena, Cartagena 1310014, Colombia
2 Facultad de Ingeniería, Universidad Tecnológica de Bolívar, Cartagena 131001, Colombia
3 Departamento de Ingeniería Hidráulica y Medio Ambiente, Universitat Politècnica de València, 46022 Valencia, Spain
4 School of Civil Engineering, University of Leeds, Leeds LS2 9JT, UK
5 Department of Civil Engineering, Architecture and Georesources, CERIS, Instituto Superior Técnico, University of Lisbon, 1049-001 Lisbon, Portugal
* Correspondence: ocoronado@utb.edu.co

Abstract: The estimation of thermodynamic behavior during filling processes with entrapped air in water pipelines is a complex task as it requires solving a system of algebraic-differential equations. A lot of different numerical methods have been used for this purpose in literature including the rigid water column (RWC) model. The main advantage of the RWC model is its acceptable accuracy with very low computational load. In that context, this research presents the computation of critical points of the physical equations that describe the phenomenon. These points provide information about the final position of the air–water interface. The Newton–Raphson method was then applied to obtain a unique equation that can be used by engineers to directly compute variables such as air pocket pressure and water column length at the end of the hydraulic event. A case study was analyzed to compare the results of the mathematical model with the obtained equation for computing critical points. Both methods provided the same values for the water column length at the end of the hydraulic event. A sensitivity analysis was conducted to identify dependent and non-dependent parameters for evaluating the critical points. The proposed formulation was validated through an experimental set of data.

Keywords: filling process; Newton–Raphson method; transient flow; pipelines; water

1. Introduction

Air–water two-phase pressurized flow is a frequently encountered phenomenon within hydraulic conduits or pipelines. It is complex to analyze since a combination of thermodynamic and hydraulic equations need to be solved. Nevertheless, this flow condition may result in catastrophic events under specific circumstances, primarily due to the amplification of unwanted pressure surges by the high compressibility of entrapped air. Such flow condition has the potential to generate a sudden, transient flow state within the system, which, in turn, can provoke rapid pressure fluctuations, characterized by exceedingly high peaks. This can occur in many industrial applications, such as in power plants, water distributions systems, water treatment facilities, and oil and gas production. An additional downside to the presence of air within a hydraulic system is its impact on the energy expenditure required for the transportation of water. When air is trapped in pipelines, it can reduce the effective cross-sectional area of flow, resulting in increased friction losses and pressure drop. This, in turn, can lead to decreased flow rates, increased energy consumption, and increased operating costs [1]. Furthermore, the presence of entrapped air pockets within water pipelines can lead to additional complications, such as unstable system operations, challenges in filter performance, a decline in pump efficiency, and pipeline corrosion. Numerous factors can lead to the entrapment or formation of air within hydraulic systems. These may include incomplete removal of air during the initial filling of the pipes, significant pressure drops that fall below the vapor pressure, malfunctions within auxiliary devices, and the operation of hydraulic machinery. For instance, a water-column separation can result in large air pockets inside hydraulic installations because of pressure drop [2–4]. Other examples of how air pockets can be introduced into pipelines are vortices in the pump intake, the operation of air vacuum valves, pipe rupture in regions of sub-atmospheric pressure, and release of dissolved air in water [5]. Despite a wealth of research, endeavors focused on developing air removal strategies to prevent the accumulation of air within hydraulic systems, a universally applicable solution remains elusive due to the highly variable nature of hydraulic systems and their operating conditions. Overall, the choice of air removal strategy will depend on the specific needs and characteristics of the pipeline system. Nevertheless, it is important to consider factors such as the size and length of the pipeline, the flow rate, and the presence of any high points or low points in the system when selecting an appropriate air removal strategy. Therefore, two-phase transient flow cannot be avoided, and water utility companies need to consider such conditions in the design and operation of water systems. From the theoretical point of view, to avoid pressure surges generated by the rapid compression of trapped air pockets during filling operations, air pockets should be continuously expelled, usually using air release valves (ARVs) [4]. Practically this can be challenging, due to common operational issues in ARVs; hence, adequate maintenance should be performed on ARVs to ensure proper operation. Furthermore, ARVs must be positioned properly to prevent pressure surges and ensure efficient air removal. For example, ARVs should be positioned at the highest point of the pipe profile where air is likely to accumulate. Pipe filling should be performed considering technical recommendations, such as the slow maneuvers of discharge valves, a pressure differential of 13.79 kPa, and considering a water flow equal to the air flow rate using a water pipe velocity around 3 m/s [5]. The most severe scenario concerning extreme pressure surges during pipe filling occurs when air release valves (ARVs) fail to function, either due to their absence or unsatisfactory maintenance. Numerous researchers have experimentally and numerically validated formulations for simulating pipe filling processes, incorporating either an elastic column model or a rigid water approach within 1D models, for systems with or without air expulsion [4,6–9]. Both types of models are used to predict pressure surges and other hydraulic transient events that may occur in the system. The choice of model depends on the specific hydraulic system and the level of accuracy required. In the majority of cases, these numerical models yield comparable results, given that the elasticity of air is significantly greater than that of water and pipe material. Furthermore, 2D and 3D Computational Fluid Dynamics (CFD) models have been utilized in recent times to simulate water filling operations in pipelines, with the aim of examining the separation of the water column, formation of geyser, displacement of manhole covers, and maximum pressure surges that are reached, among other factors [10–13]. Nonetheless, owing to their relative simplicity, 1D models are generally more convenient to utilize when compared to 2D and 3D models. As a result, the majority of the current research is directed towards refining their efficacy, covering aspects such as alternative calculation methods, examination of the unsteady friction factor, analysis of optimal orifice size for ARVs, and other relevant factors [14–17].

As a result, the numerical resolution of algebraic and differential equations for 1D models should be performed to understand the behavior of hydraulic and thermodynamic variables during the process. The calculation of the final air–water interface location is important in practical applications as well, as it provides information regarding the air pocket pressure and water column length at the end of a pipe filling, where water velocity tends to be null. An implicit formulation for calculation of the air–water interface for emptying process has been proposed [18]; however, a similar formulation is missing for filling process.

Following those lines, this research focuses on the application of the Newton–Raphson method to compute the final position of the air–water interface that occurs in a pipe filling scenario without expelled air. The Newton–Raphson method was utilized because is at least quadratically convergent. It has been applied in many applications in the hydraulic field [19]. A flowchart is presented for computing air pocket pressure and water column length at the end of a transient event using an explicit formulation, without solving the numerical resolution of the algebraic-differential equations system. A sensitivity analysis was conducted to identify parameters that can affect the final location of the air–water interface in a pipe filling, such as the polytropic coefficient, pipe slope, air pocket size, and the initial absolute pressure supplied by an energy source. In addition, a detailed velocity field analysis was performed for a case study. Finally, the developed methodology was checked using six experimental measurements in an experimental facility with a total length of 7.36 m and a nominal pipe diameter of 63 mm. Results showed that the proposed formulation provided is suitable not only for the experimental measurements but also for the analyzed case study.

2. Numerical Approach

2.1. Filling Operation

Filling processes are operations that water utility companies perform frequently in water supply networks. Water filling columns start to come into hydraulic installations while entrapped air columns are compressing, then pressure surges are achieved. The system is supplied by an energy source (tanks or pumps), which is named as p_0^*. A regulating valve (RV) is installed at the upstream end to inject water flows. Figure 1 shows a configuration of a water filling process, which is an inclined pipeline with a longitudinal slope of θ. The entrapped air pocket is at atmospheric condition (p_{atm}^*) when the hydraulic system is at rest (t = 0). During this time, the system has a null water velocity (see Figure 1a). After that, the RV is opened to fill the single pipeline, and the air pocket compression begins with a water velocity (v) and a water column length (L). The air pocket pressure (p_1^*) is chaging over time (see Figure 1b).

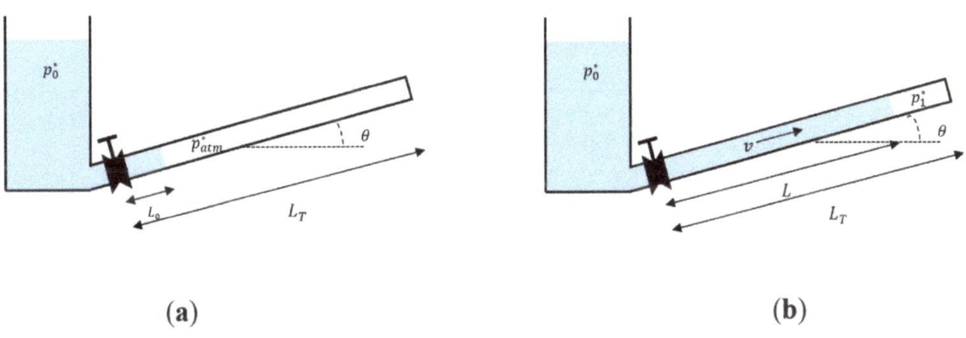

Figure 1. A filling process configuration: (**a**) hydraulic system at rest (t = 0); and (**b**) hydraulic system at $t \neq 0$.

2.2. Governing Equations

Under the assumptions of the rigid water model approach with a resistance coefficient, the filling water process is modelled as follows [3]:

1. Mass oscillation equation: Equation (1) represents the behavior of a water column along an entire pipe system [4,20].

$$\frac{dv}{dt} = \frac{p_0^* - p_1^*}{\rho L} + g \sin\theta - \frac{f}{2D}v|v| - \frac{R_v g A^2}{L}v|v| \quad (1)$$

2. Air–water interface position is given by:

$$\frac{dL}{dt} = v \tag{2}$$

3. Polytropic model: this formulation describes how the air pocket size changes with air pocket pressure pulses. These changes are described by the polytropic law (see Equation (3)), which is applied without injected air into hydraulic installations.

$$p_1^* = \frac{p_{1,0}^* x_0^k}{(L_T - L)^k} \tag{3}$$

where v = water filling velocity (m/s); p_0^* = initial pressure supplied by a tank or pump (Pa); p_1^* = air pocket pressure (Pa); ρ = water density for a specific environmental temperature; θ = pipe slope (rad), g = gravitational acceleration (m/s^2); f = friction factor; and D = internal pipe diameter; L = length of a water filling column; R_v = resistance coefficient of a valve (s^2/m^5); and A = cross-sectional area of a single pipe (m^2).

Equations (1)–(3) can be rewritten to describe the 1D movement of a water column inside the pipe as follows:

$$\frac{dL}{dt} = v \tag{4}$$

$$\frac{dv}{dt} = \frac{p_0^*}{\rho L} - \frac{p_{1,0}^* x_0^k}{\rho L (L_T - L)^k} + g \sin\theta - \frac{f}{2D} v|v| - \frac{R_v g A^2}{L} v|v| \tag{5}$$

2.3. Initial Conditions

The initial conditions of a filling water process start at rest. It implies that $v(0) = 0$; $L(0) = L_0$; $p_1^*(0) = p_{atm}^*$; $x(0) = L_T - L(0) = x_0$.

2.4. Velocity Field

To compute the analysis of velocity field, then $X(t) = (L(t), v(t))$, where the variable X render the position at time t of a particle at point (L, v) in a cartesian plane Lv . Then $X'(t) = \left(\frac{dL}{dt}, \frac{dv}{dt}\right)$ is the vector velocity of a particle at time t. However, water velocity does not depend on t, since it is a function of the position (L, v). In addition, $X'(t)$ is an autonomous system, thus the critical point of X corresponds to the final location of an air–water interface of a filling water operation. The vector field $V(L, v)$ given by the right side of the Equations (4) and (5) is expressed as follows:

$$V(L, v) = \begin{pmatrix} v \\ \frac{p_0^*}{\rho L} - \frac{p_{1,0}^* x_0^k}{\rho L (L_T - L)^k} + g \sin\theta - \frac{f}{2D} v|v| - \frac{R_v g A^2}{L} v|v| \end{pmatrix}, \tag{6}$$

Equation (6) depends on L and v, then $V(L, v)$ represents the vectors with initial point at (L, v). Thus, for a filling water process it is possible to show how a point (L, v) moves through a vector field V over time, which can be computed as:

$$X'(t) = V(X(t)), \tag{7}$$

2.5. Computation of Critical Point

Critical points are presented when $X'(t) = (0,\ 0)$; this means that $\frac{dL}{dt} = 0$ and $\frac{dv}{dt} = 0$. Considering a null velocity ($v = 0$) and by plugging Equation (5) into Equation (6), then:

$$\frac{p_0^*}{\rho L_f} - \frac{p_{1,0}^* x_0^k}{\rho L_f \left(L_T - L_f\right)^k} + g \sin\theta = 0 \tag{8}$$

The numerical resolution of Equation (8) involves the utilization of an implicit solver to obtain the final position of a filling water column (L_f). An isothermal process is characterized by a slow transient event; thus, a polytropic coefficient of k = 1.0 can be used to compute the value of L_f (see Equation (9)):

$$\frac{p_0^*\left(L_T - L_f\right) - p_{1,0}^* x_0 + \rho g \sin\theta L_f\left(L_T - L_f\right)}{\rho L_f\left(L_T - L_f\right)} = 0 \tag{9}$$

Organizing some terms:

$$-\rho g \sin\theta L_f^2 + (\rho g \sin\theta L_T - p_0^*)L_f + (p_0^* L_T - p_{1,0}^* x_0) = 0 \tag{10}$$

Applying a quadratic equation for computing L_f, then:

$$L_f|_{k=1} = \frac{\rho g \sin\theta L_T - p_0^* + \sqrt{(\rho g \sin\theta L_T - p_0^*)^2 + 4\rho g \sin\theta\left(p_0^* L_T - p_{1,0}^* x_0\right)}}{2\rho g \sin\theta} \tag{11}$$

Equation (11) can be used to compute the final position of an air–water interface during a filling operation in a single pipeline considering an isothermal evolution. However, when a polytropic evolution ($1.0 < k \leq 1.4$) is detected, there is not an explicit formulation to compute the final position of a filling water column. The Newton–Raphson method [21] was applied to get the numerical resolution for computing the value of L_f for a polytropic evolution. In this sense, the function $j\left(L_{fi}\right)$ is defined as:

$$j\left(L_{fi}\right) = \frac{p_0^*}{\rho L_{fi}} - \frac{p_{1,0}^* x_0^k}{\rho L_{fi}\left(L_T - L_{fi}\right)^k} + g\sin\theta\,, \tag{12}$$

By deriving the Equation (12):

$$j'\left(L_{fi}\right) = -\frac{p_0^*}{\rho L_{fi}{}^2} + \frac{p_{1,0}^* x_0^k}{\rho}\left[\frac{1}{L_{fi}^2\left(L_T - L_{fi}\right)^k} - \frac{k}{L_{fi}\left(L_T - L_{fi}\right)^{k+1}}\right] \tag{13}$$

An iterative process is conducted to obtain the final position L_f for a polytropic evolution. $j'\left(L_{fi}\right)$ is always different than zero. This process starts using an initial value of $L_f\big|_{k=1}$ based on Equation (11). After that, the L_{fi} consecutive values are computed as follows:

$$L_{fi+1} = L_{fi} - \frac{j\left(L_{fi}\right)}{j'\left(L_{fi}\right)} \tag{14}$$

A flowchart was created to compute the final position of L_f considering different kinds of polytropic evolutions as shown in Figure 2.

The Newton–Raphson method cannot diverge for the $j\left(L_{fi}\right)$ function, as its first derivative is negative. Therefore, $j\left(L_{fi}\right)$ always exhibits a decreasing trend without any maximum or minimum points. Additionally, the second derivative is negative, when it is close to the root, as well as when the proposed seed value is close to the solution. The $j\left(L_{fi}\right)$ function intersects the x-axis only once, making the Newton–Raphson method suitable for solving this problem.

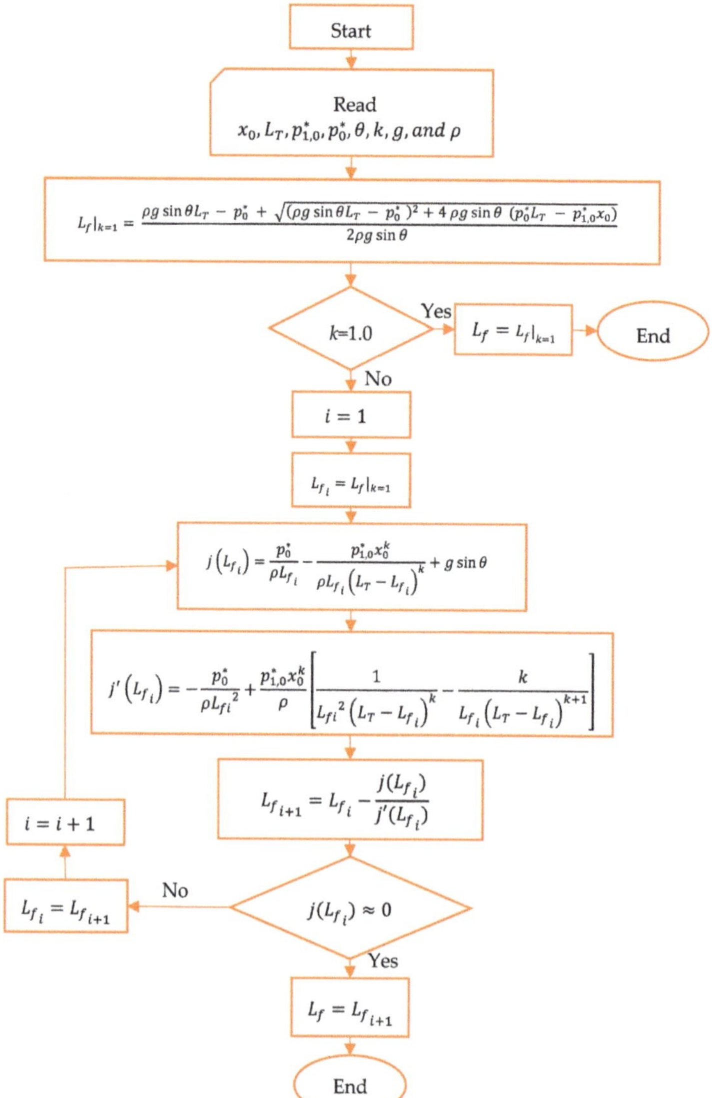

Figure 2. Flowchart for computing the final position L_f in a filling water process.

The application of the flowchart (see Figure 2) is supported based on assumptions as follows: (i) a constant friction factor, (ii) an air–water interface is modeled using a piston-flow approach, (iii) the mass oscillation equation is considered for the water phase, (iv) the polytropic law describes an entrapped air pocket evolution, and (v) the critical point theory is used to compute final positions of the hydraulic filling event.

3. Analysis of Results and Discussion

3.1. Case Study

A filling operation was simulated in a single pipe considering a total length of 600 m, an internal pipe diameter of 0.30 m, a pipe slope of 0.02 rad, a polytropic coefficient of 1.2, an initial air pocket of 500 m, a constant friction factor de 0.018, a resistance coefficient of

0.11 s^2/m^5, and an initial pressure supplied by a tank of 202,650 Pa. Figure 3 shows the evolution of the main hydraulic and thermodynamic variables over time for the analyzed case study. Figure 3a presents that the maximum water velocity occurs at 10.7 s with a value of 5.34 m/s, while the minimum value reached was −0.76 m/s (at 130.0 s), indicating a reverse water flow. After that, many oscillations occur, since the system trends to a resting position (v = 0) at the end of the hydraulic event. The pipe is filled rapidly by the water column since it takes only 130.0 s to reach a maximum value of 390.9 m. After that, the air–water interface exhibits many pulses moving around its final position (384.42 m). Figure 3b presents the evolution of air pocket pulses for the analyzed case study. The air pocket starts at atmospheric condition (101,325 Pa). The air pocket pressure head reached a peak value of 31.1 m (at 110.6 s) which was provoked by the compression of the filling water column. A similar trend is detected with regard to air pocket pressure oscillations, where the system finished with a value of 28.3 m.

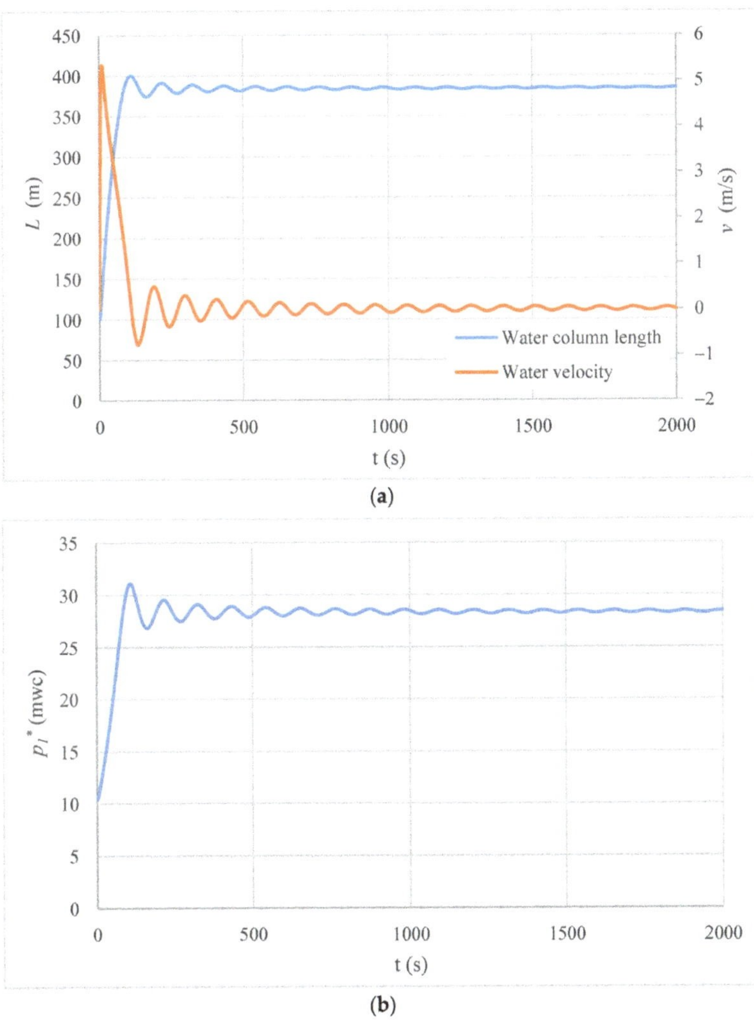

Figure 3. Evolution of hydraulic and thermodynamic variables: (**a**) water velocity and water column length; and (**b**) air pocket pressure.

It is important to highlight that there is a relationship between the water velocity and the filling water column position as shown in Figure 4. When the system is at rest ($v = 0$), the initial position of the water column length is 100 m. The end of the hydraulic event is characterized by a null water velocity as the water column length reaches its final position (384.4 m). Thus, the pipe is not completely emptied, where an air pocket size of 215.6 m is trapped at the end of the transient flow. Red points show initial and final positions of water column lengths. Orange vectors show the velocity field during the filling process occurrence. The norm of $\mathbf{V}$ shows how fast the water is moving along the pipeline.

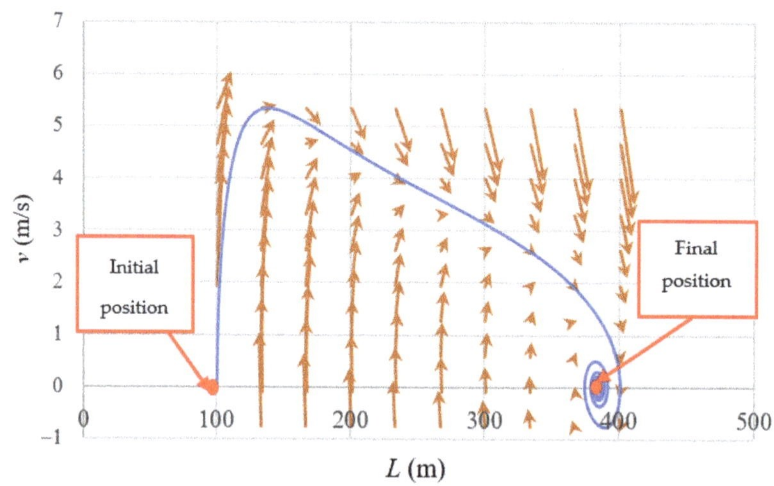

Figure 4. Relationship between water velocity and filling water column position.

3.2. Final Position of Air–Water Interface

The flowchart presented in Figure 2 contains a procedure that can be used to directly compute the final position of an air–water interface. The numerical resolution of the algebraic-differential equations system (Equations (1)–(3)) of the analyzed case study gave a final water column length of 384.42 m. The flowchart recommends starting with a seed value for an isothermal evolution since the final position can be directly computed using Equation (11), obtaining a value of 422.58 m. Table 1 shows the calculation of L_f, where the third iteration provides the required response of 384.42 m. The application of the Newton-Raphson method is suitable for predicting the final position of the air–water interface. Only three iterations were required to find the final value L_f.

Table 1. Application using the Newton–Raphson method.

i	L_i (m)	$j(y)$	$j'(y)$	L_{i+1} (m)	$L_{i+1} - L_i$ (m)
0	422.58	−0.15559	−0.00479	390.10	−32.48
1	390.10	−0.02036	−0.00365	384.53	−5.57
2	384.53	−0.00038	−0.00352	384.42	−0.11
3	384.42	0.00000	−0.00352	384.42	0.00

An analysis of Equations (1)–(3) was conducted to determine a range of seed values. The maximum value of the seed value corresponds to the total length of a water pipeline (L_T) and in all cases the final position should be greater than zero (0). Then, a sensitivity analysis of the seed value was performed considering initial values varying from 0.1 to 599 m (see Figure 5). Most selected seed values use a maximum of four iterations to find the required value (384.42 m). Many iterations are needed when selected seed values are found at the limits of physical constraints ($0 < L < L_T$), where 12 iterations were required for a seed value of 599.0 m and 15 iterations using a seed value of 0.1 m.

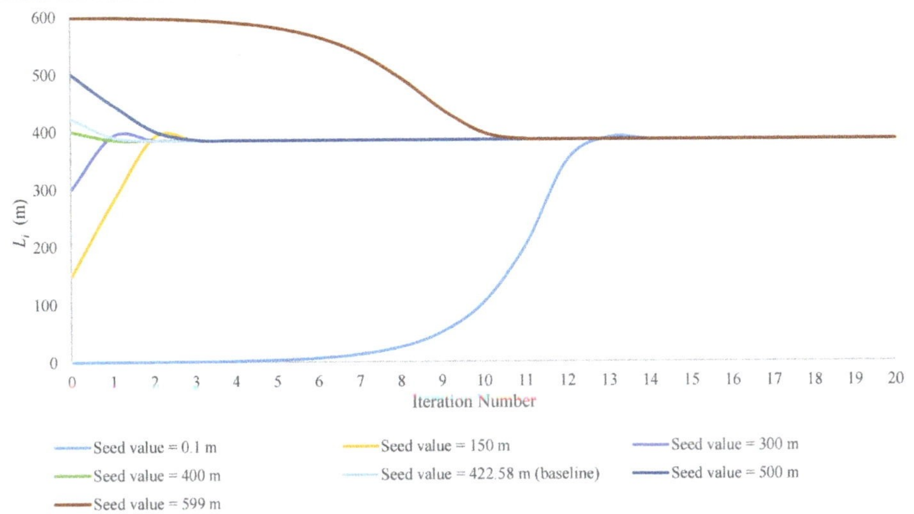

Figure 5. Sensitivity analysis of seed values.

3.3. Sensitivity Analysis

Based on data of the case study (Section 3.1), the following sensitivity analysis was conducted where only one parameter was varied in order to see the order of magnitude of the water column length and especially its final position. The analysis of Equation (8) shows that there are dependent and non-dependent parameters for computing the final position of a water column. In this sense, the final position of a water filling column (L_f) is affected by pipe slope (θ), initial air pocket size (x_0), polytropic coefficient (k), and upstream absolute pressure (p_0^*). Figure 6 shows how the value L_f can change according to the variations of these parameters. Figure 6a presents how the polytropic coefficient can affect the water velocity curve versus water column length. The lower the polytropic coefficient, the higher the final position of the obtained water column length. A value of 422.58 m is attained considering an isothermal evolution (k = 1.0), while an adiabatic behavior (k = 1.4) gives a final value of 352.96 m. Regarding the pipe slope (see Figure 6b), the lower its value, the lower the value L_f obtained. This indicates that a high pipe slope produces a more efficient filling process. The initial air pocket size is another dependent parameter that can influence the final position of an air–water interface. Small values indicate that the process will finish in a short time compared to big ones. Figure 6d shows the variation of L_f as a function of the initial absolute pressure. The smaller the initial absolute pressure supplied by a source of energy, the larger the air pocket size at the end of the hydraulic event remaining inside of the hydraulic installation. For instance, with an absolute pressure of 101,325 Pa (1 bar), the value L_f is 233.65 m, while assuming an absolute pressure of 405,300 Pa (4 bar), a value L_f of 467.11 m is achieved.

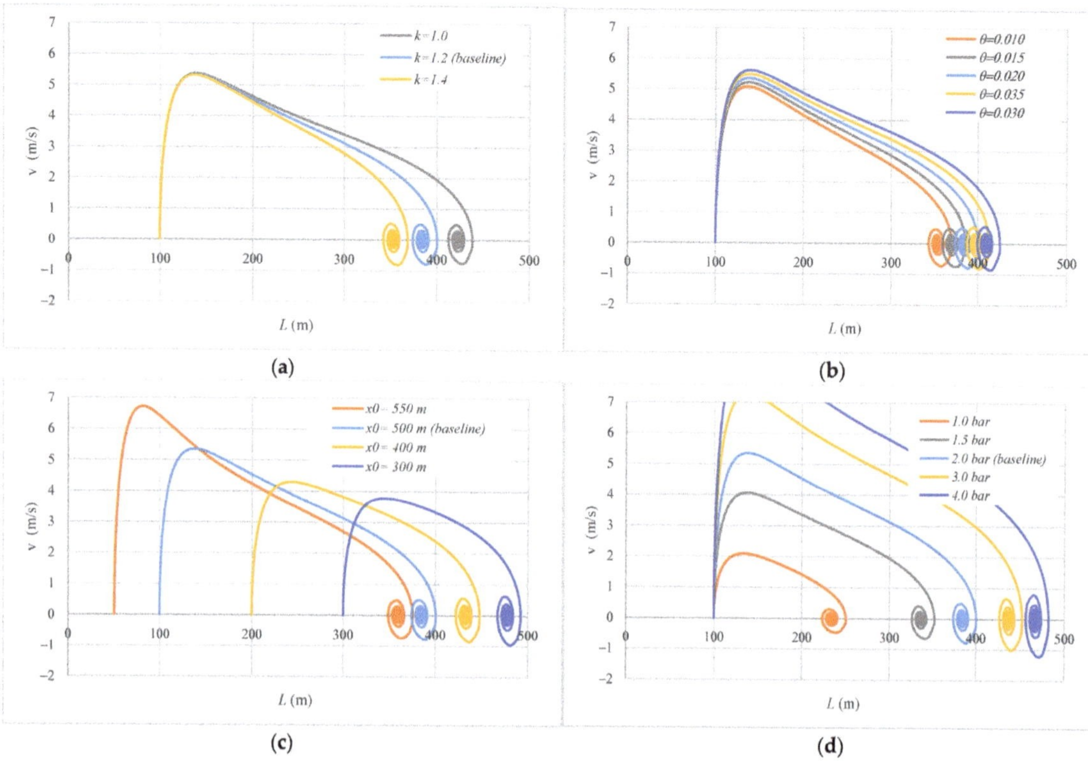

Figure 6. Sensitivity analysis for dependent parameters: (**a**) polytropic coefficient (k); (**b**) pipe slope (θ); (**c**) air pocket size (x_0); and (**d**) initial absolute pressure (p_0^*).

Figure 7 shows that internal pipe diameter, resistance coefficient, and friction factor do not affect the final position of the water column length. However, it is important to bear in mind that during the filling process, these parameters produce peak values that should be considered for design purposes. The final position of the water column length is the same independent of the variation of these parameters.

3.4. Experimental Validation

An experimental facility was set-up at the hydraulic lab at the Instituto Superior Técnico, University of Lisbon, Portugal in order to validate the methodology proposed in Figure 2. The installation is composed of a 7.36 m long PVC pipeline with a nominal diameter of 63 mm, a hydro-pneumatic tank to supply initial gauge pressures of 0.75 and 1.25 bar, and an electro-valve located at the upstream end. An entrapped air pocket was configured at the highest point of the hydraulic installation with sizes (x_0) of 0.46, 0.96, and 1.36 m, which include the configuration on the top part of the pipeline. The two inclined branch pipes have a length of 1.5 m and a longitudinal slope of 30°. The filling process starts with the opening of the electro-valve 1 (EV1). The EV2 remained closed for all experimental measurements, so the right water column did not change over time (boundary condition). Figure 8 shows the experimental facility. The left water column remained static for all experimental measurements. Table 2 presents the experimental measurements.

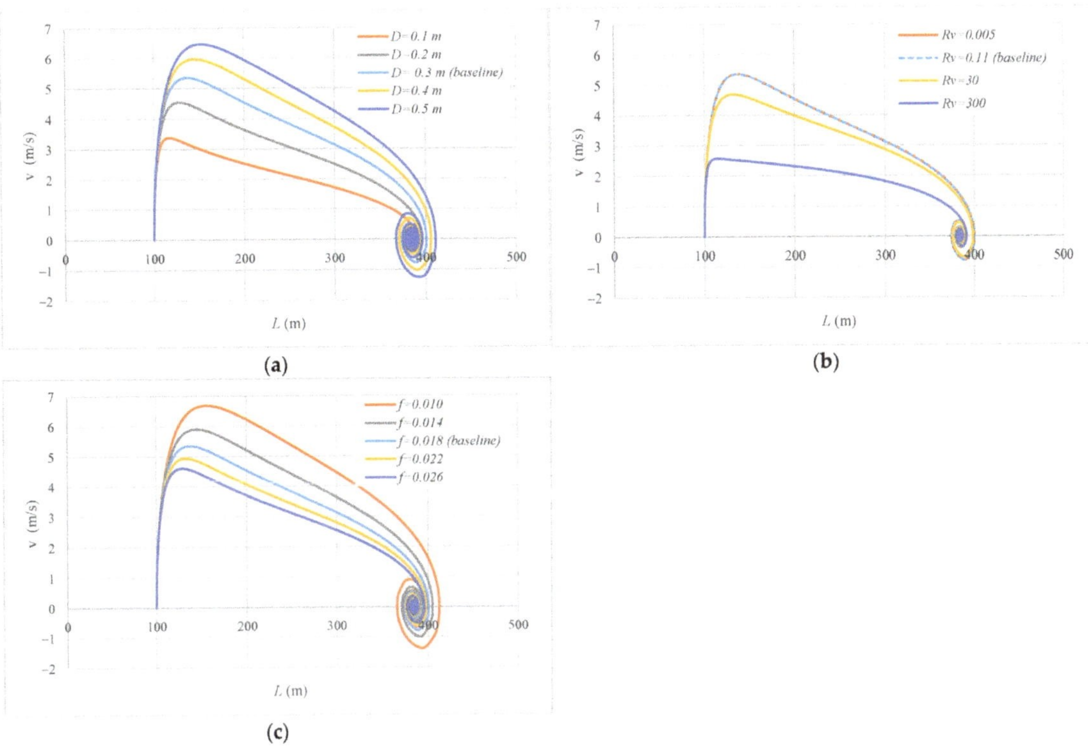

Figure 7. Sensitivity analysis for non-dependent parameters: (**a**) internal pipe diameter (D); (**b**) resistance coefficient of a regulating valve (R_v); and (**c**) friction factor (f).

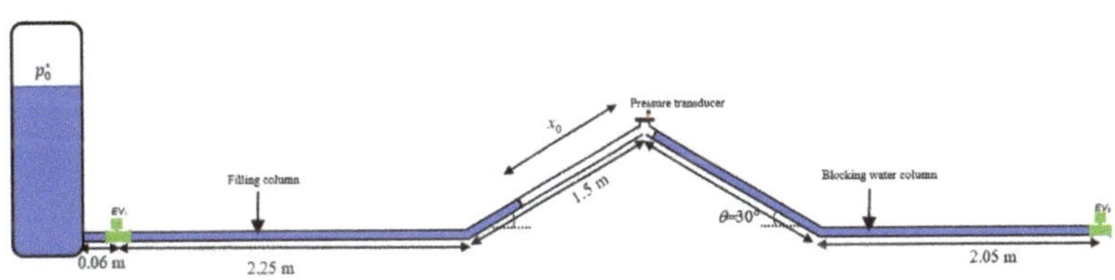

Figure 8. Experimental facility.

Table 2. Experimental measurements.

Test No.	1	2	3	4	5	6
x_0 (m)	0.46	0.96	1.36	0.46	0.96	1.36
p_0^* (bar)	1.75	1.75	1.75	2.25	2.25	2.25

A pressure transducer was installed at the highest point of the experimental facility to measure air pocket pressure. Each test was repeated twice to check the measurements. The average experimental measurements were used to validate the methodology presented in Figure 2. Figure 9a contains average absolute pressure patterns for Tests 1 to 6. The maximum value of air pocket pressure of 51.03 m (at 0.39 s) was obtained for Test No. 5, while the minimum value of air pocket pressure was obtained for Test No. 3 with a value of 31.33 m. Water column length pulses (see Figure 9b) were computed based on the polytropic law as follows:

$$L = L_T - \left(\frac{p_{1,0}^*}{p_1^*}\right)^{1/k} x_0 \tag{15}$$

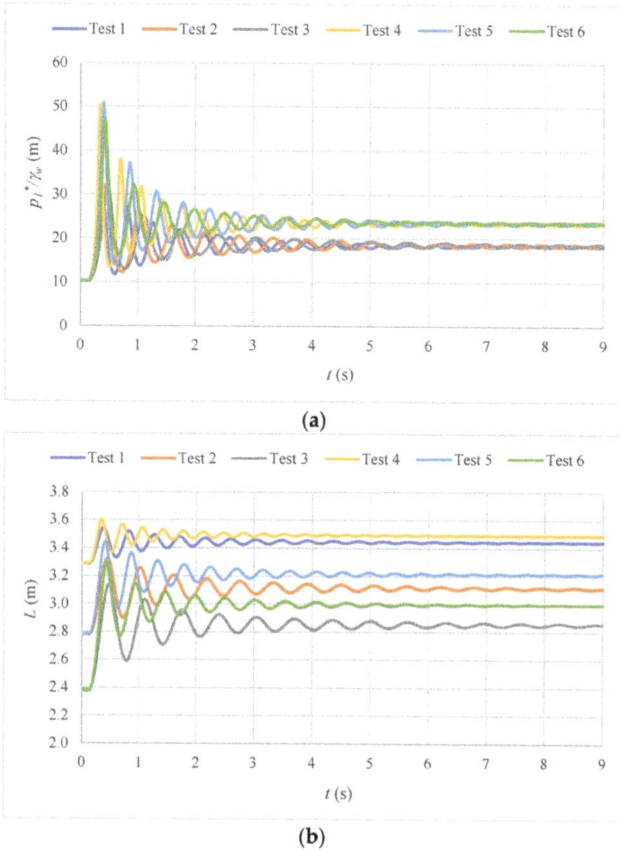

Figure 9. Evolution of the analyzed variables for the experimental facility: (**a**) air pocket pressure; and (**b**) water column lengths.

Finally, the validation of the proposed methodology was based on the final values of the water column lengths as shown in Figure 10. The comparison shows that the proposed methodology is suitable to compute the final position of an air–water interface since experimental and simulated data are similar. For instance, for Test No. 2 both experimental measurements ($L_{f,exp}$) and simulated ($L_{f,sim}$) data provided a final value of the water column length of 3.12 m. The root mean square error ($RMSE$) was computed obtaining a value of 0.017 m.

$$RMSE = \sqrt{\frac{\sum_{s=1}^{N}\left(L_{f,s,exp} - L_{f,s,sim}\right)^2}{N}} \tag{16}$$

where N is the total amount of experimental tests, and the subscript s represents a specific experimental test.

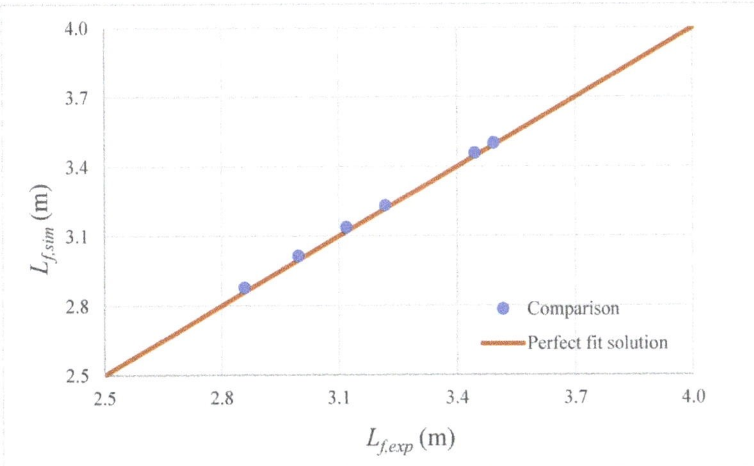

Figure 10. Comparison between experimental and simulated final water column lengths.

4. Conclusions

The analysis of filling processes with entrapped air requires the numerical resolution of an algebraic-differential equations system. However, in this research, critical points of the hydraulic system were evaluated to know the behavior of the main hydraulic and thermodynamic variables. These points represent the final position during a filling water process, where the water velocity is null, and the air pocket pressure and the water column length reach their final values. The Newton–Raphson method was applied to obtain the numerical resolution of the obtained algebraic equation. A flowchart was developed to compute the final position of an air–water interface, which contains all used equations for its calculation. The obtained equation to compute the final position of an air–water interface indicates that it depends on the polytropic coefficient, pipe slope, initial air pocket size, and absolute pressure supplied by an energy source (tank or pump). However, the internal pipe diameter, resistance coefficient, and friction factor do not alter the final location of an air–water interface. The proposed methodology was validated in an experimental facility of 7.6 m long PVC pipeline with nominal diameter of 63 mm. Results showed that experimental measurements and simulated data were very close.

Water utility companies can use the obtained formulation (see Equation (8) and Figure 1) to know the final position of an air–water interface without running the complex algebraic-differential equations system (see Equations (1)–(3)) during filling processes without an entrapped air pocket. Future research should be focused on the development of analytical solutions for computing final locations of air–water interface in rapid filling processes considering air valves.

Author Contributions: Conceptualization, Ó.E.C.-H. and D.M.B.-C.; methodology, V.S.F.-M., M.B., and H.M.R.; formal analysis, Ó.E.C.-H. and D.M.B.-C.; writing—original draft preparation, Ó.E.C.-H. and D.M.B.-C.; supervision, V.S.F.-M., M.B., and H.M.R. All authors have read and agreed to the published version of the manuscript.

Funding: This research projects did not receive either external or internal funds.

Data Availability Statement: Databases are available from the corresponding author.

Conflicts of Interest: The authors declare no conflict of interest.

Abbreviations

A	Cross-sectional area of pipe (m^2)
D	Internal pipe diameter (m)
f	Friction factor (-)
g	Gravity acceleration (m s^{-2})
k	Polytropic coefficient (-)
j	Used function in the Newton–Raphson equation
L	Water column length (m)
L_T	Pipe length (m)
p_{0^*}	Initial absolute pressure supplied by an energy source (Pa)
p_{atm}^*	Atmospheric pressure (101,325 Pa)
p^*_1	Air pocket pressure (Pa)
R_v	Resistance coefficient (ms^2m^{-6})
t	Time (s)
v	Water velocity (m s^{-1})
V	Vector field on a region in the plane (L, v)
x	Air pocket size (m)
X	Vector function of t
ρ	Water density (kg m^{-3})
θ	Pipe slope (rad)
γ_w	Water unit weight (N m^{-3})
Subscripts	
0	Refers to an initial condition
f	Refers to a final condition
i	Iteration number
s	Specific experimental test
Superscripts	
$'$	Derivative

References

1. Tasca, E.; Besharat, M.; Ramos, H.M.; Luvizotto, E., Jr.; Karney, B. Contribution of Air Management to the Energy Efficiency of Water Pipelines. *Sustainability* **2023**, *15*, 3875. [CrossRef]
2. Martins, N.M.; Soares, A.K.; Ramos, H.M.; Covas, D.I. CFD modeling of transient flow in pressurized pipes. *Comput. Fluids* **2016**, *126*, 129–140. [CrossRef]
3. Maddahian, R.; Shaygan, F.; Bucur, D.M. Developing a 1D-3D model to investigate the effect of entrapped air on pressure surge during the rapid filling of a pipe. *IOP Conf. Ser. Earth Environ. Sci.* **2021**, *774*, 012069. [CrossRef]
4. Fuertes-Miquel, V.S.; Coronado-Hernández, Ó.E.; Mora-Melia, D.; Iglesias-Rey, P.L. Hydraulic Modeling during Filling and Emptying Processes in Pressurized Pipelines: A Literature Review. *Urban Water J.* **2019**, *16*, 299–311. [CrossRef]
5. AWWA (American Water Works Association). *Manual of Water Supply Practices M51—Air Valves: Air Release, Air/Vacuum and Combination*; AWWA: Denver, CO, USA, 2016.
6. Zhou, L.; Liu, D. Experimental Investigation of Entrapped Air Pocket in a Partially Full Water Pipe. *J. Hydraul. Res.* **2013**, *51*, 469–474. [CrossRef]
7. Tijsseling, A.S.; Hou, Q.; Bozkuş, Z. Rapid Liquid Filling of a Pipe With Venting Entrapped Gas: Analytical and Numerical Solutions. *J. Press. Vessel. Technol.* **2019**, *141*, 041301. [CrossRef]
8. Liou, C.P.; Hunt, W.A. Filling of pipelines with undulating elevation profiles. *J. Hydraul. Eng.* **1996**, *122*, 534–539. [CrossRef]
9. Izquierdo, J.; Fuertes, V.; Cabrera, E.; Iglesias, P.; Garcia-Serra, J. Pipeline start-up with entrapped air. *J. Hydraul. Res.* **1999**, *37*, 579–590. [CrossRef]
10. Wang, H.; Zhou, L.; Liu, D.; Karney, B.; Wang, P.; Xia, L.; Ma, J.; Xu, C. CFD approach for column separation in water pipelines. *J. Hydraul. Eng.* **2016**, *142*, 04016036. [CrossRef]
11. Chan, S.N.; Cong, J.; Lee, J.H. 3d numerical modeling of geyser formation by release of entrapped air from horizontal pipe into vertical shaft. *J. Hydraul. Eng.* **2018**, *144*, 04017071. [CrossRef]
12. Wang, J.; Vasconcelos, J. Manhole cover displacement caused by the release of entrapped air pockets. *J. Water Manag. Model.* **2018**, *26*, 1–6. [CrossRef]
13. Martins, N.M.; Delgado, J.N.; Ramos, H.M.; Covas, D.I. Maximum transient pressures in a rapidly filling pipeline with entrapped air using a CFD model. *J. Hydraul. Res.* **2017**, *55*, 506–519. [CrossRef]
14. Tijsseling, A.; Hou, Q.; Bozku¸s, Z.; Laanearu, J. Improved One-Dimensional Models for Rapid Emptying and Filling of Pipelines. *J. Press. Vessel Technol.* **2016**, *138*, 031301. [CrossRef]

15. Zhou, L.; Cao, Y.; Karney, B.; Vasconcelos, J.; Liu, D.; Wang, P. Unsteady friction in transient vertical-pipe flow with trapped air. *J. Hydraul. Res.* **2021**, *59*, 820–834. [CrossRef]
16. Zhou, L.; Pan, T.; Wang, H.; Liu, D.; Wang, P. Rapid air expulsion through an orifice in a vertical water pipe. *J. Hydraul. Res.* **2019**, *57*, 307–317. [CrossRef]
17. Zhou, L.; Cao, Y.; Karney, B.; Bergant, A.; Tijsseling, A.S.; Liu, D.; Wang, P. Expulsion of Entrapped Air in a Rapidly Filling Horizontal Pipe. *J. Hydraul. Eng.* **2020**, *146*, 04020047. [CrossRef]
18. Coronado-Hernández, O.E.; Bonilla-Correa, D.M.; Lovo, A.; Fuertes-Miquel, V.S.; Gatica, G.; Linfati, R.; Coronado-Hernández, J.R. An Implicit Formulation for Calculating Final Conditions in Drainage Maneuvers in Pressurized Water Installations. *Water* **2022**, *14*, 3364. [CrossRef]
19. Canelon, D.J. Pivoting Strategies in the Solution of the Saint-Venant Equations. *J. Irrig. Drain. Eng.* **2009**, *135*, 96–101. [CrossRef]
20. Martin, C.S. Entrapped Air in Pipelines. In Proceedings of the Second International Conference on Pressure Surges, London, UK, 22–24 September 1976.
21. Chapra, S.; Canale, R. *Numerical Methods for Engineers*, 7th ed.; Mcgraw-Hill Education, Cop: New York, NY, USA, 2015.

water

MDPI

Article

Application of Machine Learning Coupled with Stochastic Numerical Analyses for Sizing Hybrid Surge Vessels on Low-Head Pumping Mains

Ahmed M. A. Sattar [1,2,*], Abedalkareem Nedal Ghazal [1], Mohamed Elhakeem [3], Amgad S. Elansary [1] and Bahram Gharabaghi [4]

1 Department of Irrigation and Hydraulics, Cairo University, Giza 12613, Egypt
2 Department of Civil Engineering, German University in Cairo, Cairo 11835, Egypt
3 Civil Engineering Department, Abu Dhabi University, Abu Dhabi P.O. Box 59911, United Arab Emirates
4 School of Engineering, University of Guelph, Guelph, ON N1G 2W1, Canada
* Correspondence: ahmoudy77@yahoo.com

Abstract: In surge protection, low-head profiles are deemed a challenge in pump failure events since they are prone to severe negative pressure surges that require an uneconomical surge vessel volume. A hybrid surge vessel with a dipping tube can provide required protection with reasonable economic volume. This work presents novel analyses for the hybrid surge vessel and develops a simple model for its optimum sizing using a stochastic numerical approach coupled with machine learning. Practical ranges for correct sizing of vessel components, such as ventilation tube, inlet/outlet air valves, and compression chamber, are presented for optimal protection and performance. The water hammer equations are iteratively solved using the hybrid surge vessel's revised boundary conditions within the method of characteristics numerical framework to generate 2000 cases representing real pump failures on low-head pipelines. Genetic programming is utilized to develop simple relations for prediction of the hybrid vessel initial and expanded air volumes in addition to the compression chamber volume. Moreover, the developed model presented a classification index for low-head pipelines on which the hybrid vessel would be most economical. The developed model yielded good prediction error statistics. The developed model proves to be more accurate and easier to use than the classical design charts for the low-head pumping mains. The model clearly showed the relation between various hydraulic and pipe parameters, with pipe diameter and static head as the most influencing parameters on compression chamber volume and expanded air volume. The developed model, together with the classification indices, can be used for preliminary surge protection sizing for low-head pipelines.

Keywords: water hammer; transient; machine learning; stochastic numerical analysis; flat profile

1. Introduction

When a pump trips or loses power, a negative pressure wave is generated at the pump. Cavitation and water column separation may occur if the negative pressure wave falls below the water vapor pressure, which is 10 m below atmospheric pressure at 20 °C. Water columns oscillate as a result of the separation, and they can reassemble when air bubbles collapse and create excess pressure in the pipeline. This excess pressure can cause severe damage to the water supply system and economic losses. Thus, protection of water supply systems from transient events, positive and negative, has been the focus of much research and practical studies.

Since the transient wave is greatly influenced by the pipeline material, the replacement of metallic pipes in hydropower plants and water supply systems with viscoelastic material has begun [1]. These viscoelastic pipes would suppress the pressure transient waves due to low impedance and fast damping when compared with rigid metallic pipes.

Metallic-polymeric pipes were proposed by [2] to upgrade existing steel pipes in negative transient events. Moreover, for upgrading the existing steel pipe network, ref. [3] proposed the replacement of a short section of steel pipe with branching plastic pipe to avoid major modifications. It has also been reported that glass reinforced plastic pipe was more effective in control of negative transients when compared to high-density polyethylene [1]. Recently, ref. [4] proposed a combination of glass reinforced pipe and steel pipe for controlling negative transients. However, most of this work concentrated on hydropower plant conduits, and reported shortcomings such as requirements of major modifications in existing systems to accommodate new plastic pipes, and manufacturing limitations of pressure and size of plastic pipes [4], in addition to ineffectively eliminating the initial generated negative surge wave generated by pump trip.

Many devices have been developed and successfully used to protect pipelines from hydraulic transients, e.g., surge vessels, one-way surge tanks, air valves, and relief valves. Of these, air valves, one-way surge tanks, and surge vessels are effectively used to control negative transients in pumping mains [5]. For a water intake pumping station, the protective effect against negative transients was studied and confirmed by [6] following a pump trip using an air valve on a discharge line of pump. Using a combination of air valves, ref. [7] showed that long pumped pipelines can be effectively protected against negative transients using an air valve in addition to two-staged close butterfly valves. On the other hand, ref. [8] showed that negative transients can be controlled in a cascading pumping station by using a system of air vent valves along the pipelines for prevention of vacuum cavities. Using a pressure relief valve and safety membrane has been shown by [9] to effectively reduce positive transients and, to a lesser extent, limit the negative transients in a small hydropower plant. The operation of air valves following pump trip transients have been studied by [10], where they stated that two-step closure of air valves would reduce negative transients compared to one-step valve closure. However, calculation of air valve size and location is extremely sensitive and, if not correctly sized and placed on pumping mains, can lead to destructive effects during negative transients [9]. Moreover, the usage of air valves only on pumping mains for protection against negative transients can fail to reach the required efficiency especially for large-diameter pipes [11].

To allow for larger air volumes to enter the pipeline following pump trip transient, the one-way surge tank can be used. It would effectively control negative transients by allowing atmospheric air to enter into the tank and push more water to enter the pipeline during negative transient wave, but would need a large height to prevent water spillage outside the tank during the positive transient wave making it an uneconomic solution. For a deep well pump group, ref. [12] studied the effect of one-way surge tank on pump trip negative transients and found that a combination of one-way surge tank and multiple air valves would be needed and would only be effective in small flow systems [12].

Surge vessels can be the efficient solution to protect pipelines from these transients. They can effectively alleviate negative transients in addition to reduction of positive transients in pumping mains [5]. Ref. [13] studied transients generated on a rising pump main due to pump failure. They used the rigid column theory and developed design charts for initial air volume of surge vessels. Adding differential orifice plates to the surge vessel, ref. [14] proposed design charts for initial air volume with reduced size compared to those provided by [13]. Other simpler charts for initial air volume were developed by [15] for the surge vessel on a rising pump main considering various diameters of the inlet and outlet tank connecting pipe. With the help of the artificial neural network, ref. [16] collected data from [13] charts and developed a model for the surge vessel initial air volume. The range of parameters used were not reported and surge vessel throttling was not included. Two approximate expressions for the calculation of pipeline pressure for quick sizing of surge vessel volume and orifice resistance were presented by [17]. However, all these models have many shortcomings, making them not applicable for even preliminary stages of surge vessel design [17]. They are based on the rigid column theory's simplifying assumption and the De Sparre rule, which restricts their applicability. Many models have not considered pipe friction, which can impact the final size of the surge vessel. Furthermore, only the initial air volume could be predicted from any model; none of them offered the final expanded air volume. To optimize design of the surge vessel, ref. [18] used the sequential quadratic programming method for solution of transients following a pump trip. The optimization algorithms have been further developed and used by many researchers, compiled with numerical analyses to optimize surge vessel volume [19] and installation position on pipeline [20]. While there are many tank variables and hydraulic operating conditions, the optimization-numerical approach involves lots of calculations and is case specific, with no general formula to be applied easily in the planning phases of the project. Based on the incompressible flow theory and simple harmonic vibration model, an approximate formula for the optimal size of the air vessel was presented [21]. The formula did not include the pipeline diameter, despite being an important influential parameter on the size of the surge vessel, and the formula was tested on only one test case. Using evolutionary algorithms, a simple model was developed for the prediction of initial and final air volumes in a surge tank considering various pipe and operational conditions [22]. The developed formulae are easy to apply and have been developed based on thousands of generated numerical pump failure cases. Based on the Krylov–Mitropolsky method, an analytical mode for the water level oscillations in the surge vessel was developed [23]. They used the developed model for optimal volume selection of the air vessel and used two practical cases for testing. An intelligent self-controlled surge tank was developed, including a damper to reduce required volume of normal surge tank [24]. The damper adjusted flow depending on the differential pressure on both sides of the connector; this helped more flow volume to be pushed out of the vessel at higher rates in case of pump trips. In addition to the importance of surge vessel size in protection against pump tri negative transients, the location of the surge vessel on the pump main was reported to have an influence on the pressure waves' amplitude [21].

In addition to the above-mentioned shortcomings, none of the available models has addressed low-head pumping profiles. In surge protection design, a low-head profile is deemed as a challenge for hydraulic engineers especially in pump failure events. A low static head makes the pipeline prone to more severe negative pressure surges that require a quick supply of water from the surge vessel to compensate for the generated sub-atmospheric pressure regions. This means that the initial compressed air volume in the surge tank has to be substantially more than in vessels installed on steep profiles so that when expanded, it will deliver the required volume of water with adequate energy. These models produce huge surge vessels with uneconomical volumes on low-head pumping mains, reaching 1000 cubic meters in some cases. Bladder surge vessels have been proposed to provide quite economical surge vessels for these low-head profiles. However, the vessel did not provide the required economical volumes (even if they are less than standard surge vessels) in addition to the imposed risk of connection pipe obstruction when used in sewage systems [22].

Charlatte [25] filed a patent for a self-regulating water hammer protection wastewater tank (ARAA). The new tank has a central ventilation tube or dipping tube, providing more economical vessel volumes and requiring less maintenance than bladder vessels. This hybrid surge vessel combines the principle of operations of a standard surge vessel, an open-vented surge vessel, and an air valve during the operation cycle and thus is called, in this work, a hybrid vessel. The general principles and working scheme of the AARA-vessel was presented as an effective solution for the negative surges following pump trip [26]. They provided a couple of practical field applications to support their study analyses. Ref. [27] presented design charts for surge pressure rise in pipe lines with pump failure and check valve closure cases. They were concerned about the process rather than the surge device installation and only studied the effect of friction and pump characteristics on the surge pressure. However, they were the first to provide charts that are directed only to low head pump lines, though they did not mention a clear description or physical range for a low head line. When presenting guidelines for choosing the transient control devices, the hybrid vessel with a dipping tube was included and a brief description was given in [28]. However, no further analysis was performed on the dipping tube tank in their work and no guidelines for where to use such device was provided. A surge study was performed for a 250 km pipeline, diameter 1.6 m, from the Shuweihat desalination plant [29]. They reported that the pipeline profile was flat with few local high points, and that this resulted in excessive negative transients. To control them, Deltares presented an innovative vertical air vessel equipped with a passive air release valve installed on the air vessel. It was referred to as a hybrid air vessel and a hydraulic model was described; a cost saving of 50% was reported using the hybrid vessel. They were the only study that mentioned that normal surge vessels would be not economic for flat profiles, but they did not give any clear numerical model for the vessel, nor did they define a definition of flat profiles, where the hybrid vessel would be useful. Refs. [30,31] stated that the hybrid vessel with a dipping tube would be the most efficient solution for protection of long pipe lines following a pump failure. They presented a simple, yet comprehensive, internal boundary conditions for the dipping tube vessel based on hydraulic characteristics of the dipping tube and how they would interact with existing equations for surge vessel. Their model was developed in the context of the Method of Characteristics and could enable the correct sizing of the dipping tube vessel. A combination of air chamber and air-inlet valves is studied to optimize transient protection [32]. This combined method obtained the best protective scheme where transient pressures are maintained in a safe standard while minimizing the protection cost. Recently, an air chamber with an adjustable standpipe tank was proposed [33]. A simple numerical study was performed, and a new tank was compared to surge vessel, but the study reported the need for a larger volume novel tank when compared to a normal surge vessel and recommended more research work on tank parameters and their impact on surge protection.

While a considerable amount of research work for sizing surge vessels and other protection devices on pumping mains has been made, the hybrid vessel with a dipping tube lacks similar research. To the authors' knowledge, no research has addressed the sizing of the hybrid vessel with a dipping tube. There are no available practical ranges for the hybrid surge vessel initial gas volume, compression chamber volume, and ventilation tube length/diameter. There is no clear definition and limiting range of pump main physical and operational parameters that define a low-head profile, where the hybrid vessel with dipping tube would be best suited for. This would be beneficial to hydraulic practical engineers, especially because the hybrid vessel with a dipping tube has been included in some commercial water network design software [34].

The primary goal of this work is to contribute to the literature a hybrid surge vessel with a dipping tube on pump mains. In addition, this study seeks to create a new, rough approximation model that will allow for the prediction of the ideal hybrid surge vessel size for the protection of low-head pumping mains in the event of a pump failure. This will be achieved by: (i) providing in-depth analysis on the operation concept of the hybrid surge vessel in protection of pumping mains against pump failures, (ii) providing practical ranges for low-head profiles where a hybrid surge vessel would be more economical to use than standard surge vessels, (iii) providing practical ranges for correct sizing of vessel components, such as ventilation tube, inlet/outlet air valves, and compression chamber, and (iv) developing simple straight forward relations that can calculate the initial air volume, expanded air volume, and the compression chamber volume for the hybrid vessel. The developed model enables the practical usage of hybrid surge vessels on low-head pumping mains and thus overcomes the economic limitations of using standard surge vessels. Moreover, they would be easily adopted by non-experienced users for preliminary choice and sizing of hybrid vessels when compared to complex commercial software that need many iterations and experienced users.

To develop model, random inputs covering a variety of pumping scenarios are generated using Monte Carlo simulation (MCS) as a tool for stochastic analysis in combination with the numerical solution using the Method of Characteristics. Prediction models for the initial air volume, final tank volume, and compression chamber volume are developed using genetic programming (GP). Statistics are used to evaluate and compare the generated genetic programming models to the classical existing design charts. Error analysis is carried out to verify the predicted uncertainties with additional quantification. In addition, parametric analysis is used to verify the physical trends of new model predictions.

Hybrid Surge Vessels

In a pump failure, the surge vessel is regarded as the most effective method of controlling created transients. It can prevent negative transients and minimize the effect of positive transients. When a pump fails, the hydraulic grade line (HGL) drops suddenly below the pipe centerline, creating a region in the liquid with sub-atmospheric pressure. Starting at the upstream pipeline, this created negative pressure wave moves continuously downstream until dampened. The initial negative peak can result in the worst hydraulic condition for the pipe. with the formation of vapor cavities leading to column separations in case pressure drops to the liquid vapor pressure. To compensate for the negative pressure transient and to maintain the positive momentum of the water flowing downstream of the pipe, compressed air in the surge vessel pushes the water out to the pipeline. In addition to the pipeline friction and initial static head, the action of the surge vessel also reduces additional pressure oscillations. The initial drop in the hydraulic grade line following pump failure will be more severe for low-head pipeline profiles than in typical rising profiles because of the low static head, which increases the likelihood of column separation. Due to the low static head, the generated negative pressure wave is only damped by the pipeline friction with no resistance from the static head, leading to faster prolonged negative pressure oscillations. Using a surge vessel in this situation would require a very large vessel volume in order to provide the adequate amount of water to compensate for the pressure loss at the pump location and continuously decelerate the returning pressure wave. When water is forced into the pipeline, the vessel volume must be large enough to hold the full expanded air volume. Therefore, a hybrid surge vessel is recommended to be used in this situation for better performance and reasonably reduced vessel volume. The hybrid surge vessel with a ventilation tube is shown in Figure 1. The tank has a height of H_{tank} and a diameter of D_{tank}. The central tube is used for ventilation and is termed, in the following work, as a dipping tube or ventilation tube. The ventilation tube has the height and diameter of H_{dt} and D_{dt}, respectively. The total vessel volume, $\forall_{tank}$, is equal to the initial compressed air volume, $\forall^{o}_{air}$, in addition to the water volume below it. The compression chamber is the part of the vessel surrounding the central ventilation tube having the same diameter as the vessel and the height of the ventilation tube, and volume $\forall_{cc}$ is equal to $0.25\pi\left(D^2_{tank} - D^2_{dt}\right)H_{dt}$. Air entry and exit to/from the ventilation tube is conducted through inlet and outlet air valves, as shown in the figure. The vacuum valve is equipped with a check valve to allow air only in the ventilation tube, while the release valve is smaller in diameter to let air out more slowly [30].

This is called a hybrid vessel since it combines the principles of operation of various protection devices during its operation cycle. When air is trapped in the compression chamber, it functions as a regular surge vessel, a vented surge vessel when the ventilation pipe's float is open, and an air valve when all the water has been forced into the pipeline. When the pump main starts to fill, water enters the hybrid vessel, expelling air from the ventilation pipe while the float is still open. The level of water increases in the hybrid vessel until it reaches the bottom of the ventilation tube. Then, it continues to increase in the central ventilation pipe as the upper float is still open to the atmosphere. The air in the compression chamber is being compressed. When the rising water pushes the float to the upper limit of the central ventilation pipe, the shutoff valve is closed while the air in the compression chamber is being compressed until the system is balanced and water ceases to enter the hybrid vessel. Following pump failure, the compressed air at the top of the hybrid vessel pushes water to the pipeline until the elastic energy of the compressed gas is depleted. This is when all the water stored above the bottom of the ventilation tube has been delivered to the pipeline and the ventilation tube starts to empty and float open. From this instant, the vessel is opened to the atmosphere and water is delivered to the pipeline under constant atmospheric pressure. The water level continues to decrease until a low level is reached when the reverse pressure wave reaches the location of the vessel. The initial hydraulic grade line is reached following the damping of pressure oscillations by the cushioning effect in the hybrid vessel and pipe friction. Due to the recent application of the hybrid surge vessel in practical projects, there are no available practical ranges for the initial gas volume, compression chamber volume, and ventilation tube length/diameter.

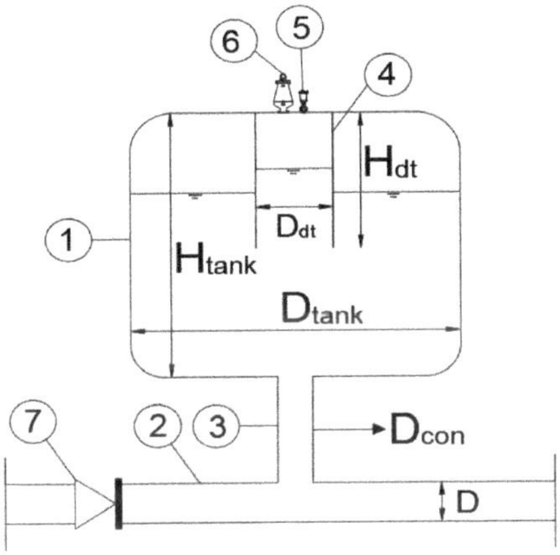

Figure 1. A schematic of a typical hybrid surge vessel for protection of low head pipelines (1-Hybrid surge vessel, 2-water main, 3-hybrid vessel inlet pipe, 4-dipping tube, 5-air release valve, 6-vacuum valve, 7-check valve).

2. Materials and Methods

2.1. Allowable Safe Pressure Limits

The maximum positive H_{max} pressure surge has no generally accepted upper limit, notably in countries including the USA and England. Engineering standards and pipe material rating play a large role in regulating this restriction. However, some scholars proposed that the zero gauge pressure head limits the minimum pressure H_{min} surge; others reported negative gauge pressures of −7 m, e.g., ref. [22]. Practice, however, showed that compared to the scenario of zero pressure head, a −3 m pressure head led to a significant reduction in vessel size with no impact on pipe integrity. To avoid the worst-case situation for the pipe, the surge vessel size is determined in this work to keep H_{max} 40% above the steady-state hydraulic grade and to keep the pipe's rated pressure from being exceeded. On the other hand, H_{min} is restricted to a value of −3 m. These restrictions can be verified by looking at the calculated pressure profiles along the pipeline following a pump failure. The calculation of these transient profiles is briefly explained in the following section.

2.2. Numerical Modelling of Pressure Profile along the Pipeline

The one-dimensional continuity and momentum equations describing unsteady ideal flow in pipes are written as:

$$\frac{\partial H}{\partial x} + \frac{1}{gA}\frac{\partial Q}{\partial t} = 0 \quad (1)$$

$$\frac{\partial Q}{\partial x} + \frac{gA}{a^2}\frac{\partial H}{\partial t} = 0 \quad (2)$$

where x = the distance along the pipe measured positive in the downstream direction (m); t = time step of solution (s); H = the piezometric head (m); Q = water flow in pipeline

(m^3/s); a = the wave speed (m/s); A = cross-sectional area of pipeline (m^2); and g = the gravitational acceleration (m/s^2).

Using the Method of Characteristics (MOC), the one-dimensional continuity and momentum equations for pipe transient flow can be written as follows along the +ve and −ve characteristics lines, respectively:

$$Q_b^{t+\Delta t} + C_a H_b^{t+\Delta t} = Q_a^t + C_a H_a^t - R\Delta t (Q_a |Q_a|)^t \tag{3}$$

$$Q_a^{t+\Delta t} - C_a H_a^{t+\Delta t} = Q_b^t - C_a H_b^t - R\Delta t (Q_b |Q_b|)^t \tag{4}$$

where a, b are any two arbitrary consecutive points along the solution grid along the pipe length; $C_a = g \times A/a$; t denotes the time step at current values (s); while $t + \Delta t$ denotes the following time step (s); $R = f/2DA$; D = the pipe diameter (m); and f is the Darcy–Weisbach friction coefficient.

Modeling the hybrid surge vessel requires including the ventilation tube and the compression chamber in numerical simulations. The numerical model for the hybrid surge vessel was developed by [30,31]; equation from this study shall be adopted in this work. The continuity equation inside the hybrid vessel can be described as the summation of flows of the dipping tube and the compression chamber:

$$Q_s = Q_{dt} + Q_{cc} \tag{5}$$

where Q_{cc} = water flow inside hybrid vessel to/from compression chamber (m^3/s); Q_{dt} = water flow inside hybrid vessel to/from dipping tube (m^3/s); and Q_s = water flow to and from the hybrid vessel (m^3/s). The air pressure inside the hybrid vessel can be written as:

$$H_{air} = H + H_{bar} - Z_s - h_{loss} \tag{6}$$

H_{air} = air pressure inside the hybrid surge vessel (m); Z_s = the water height in the vessel (m); H_{bar} = atmospheric head (m); H = the piezometric head inside hybrid vessel (m); and h_{loss} = losses of flow through vessel connecting pipe (m), and defined as $h_{loss} = C_{loss} Q_s |Q_s|$, and C_{loss} = loss coefficient. The air mass flow rate inside the hybrid vessel is written as [30]:

$$Q_m = 0.6 \rho_o A_{or} v_{max} \left(1 - (H_{air}/H_{bar})^{1-1/n}\right)^{0.5} \tag{7}$$

where Q_m = mass flow rate of air inside hybrid tank (kg/s); ρ_o = air density at atmospheric pressure (kg/m^3); A_{or} = the cross-sectional area of the air in/out flow orifice (m^2); n = polytropic gas equation exponent; and v_{max} = the maximum air velocity through air orifice and defined as [30]:

$$v_{max} = (2nH_{air}/(n-1)\rho_o)^{0.5} \tag{8}$$

The variations in the air volume $\forall_{air}$ in the hybrid vessel during the transient event can be written as follows:

$$\forall_{air}^{t+\Delta t} = \forall_{air}^t - 0.5\Delta t \left(Q_s^{t+\Delta t} - Q_s^t\right) \tag{9}$$

where $\forall_{air}$ = volume of the air enclosed in the hybrid surge vessel (m^3); and Δt = solution time step (sec.).

Modeling the behavior of the hybrid surge vessel requires the knowledge of the water level inside the vessel concerning the ventilation tube bottom level, Z_{dp} [30]. When the water level is in the middle of the ventilation tube, Q_s can be calculated as follows:

$$Q_s^{t+\Delta t} = \frac{2}{\Delta t}\left(\forall_{air\,dt}^t + \forall_{air\,cc}^t - \left(\frac{H_{bar}}{H_{air\;cc}^{t+\Delta t} - \frac{\forall_{air\,cc}^t}{A_{tank} - A_{dt}} + \frac{\forall_{air\,dt}^t}{A_{dt}}}\right)^{1/n} \frac{m_{dt}^t - 0.5\Delta t\left(Q_m^{t+\Delta t} - Q_m^t\right)_{dt}}{\rho_o} - \left(\frac{H_{bar}}{H_{air\;cc}^{t+\Delta t}}\right)^{1/n} \forall_{air\;cc}^{t+\Delta t}\right) - Q_s^t \tag{10}$$

Otherwise, when the water level is below the ventilation tube bottom, Q_s can be calculated as follows:

$$Q_s^{t+\Delta t} = \frac{2}{\Delta t}\left(\forall_{airtank}^{t} - \left(\frac{H_{bar}}{H_{air}^{t+\Delta t}}\right)^{1/n} \frac{m^t - 0.5\Delta t\left(Q_m^{t+\Delta t} - Q_m^t\right)}{\rho_o}\right) - Q_s^t \tag{11}$$

where A_{tank} = the cross-sectional area of the hybrid tank (m^2); A_{dt} = cross-sectional area of the dipping tube (m^2); ρ_o = air density at atmospheric pressure (kg/m^3); m = mass of air inside the hybrid surge vessel (kg); m_{dt}^t = mass of air inside the dipping tube (kg); $Q_m^{t+\Delta t}$ = mass flow rate of air inside hybrid tank (kg/s); and n = polytropic gas equation exponent with value of 1.2 assuming the contained air in the tank satisfies the polytropic equation for a perfect gas.

The hybrid surge vessel is connected to the pipeline downstream pump discharge. Since the pipe on the suction side of the pump is expected to be short, it is not considered in the study. The hybrid surge vessel is modelled, inside the method of characteristics, as an internal boundary condition with the following equations at the location of the vessel:

$$Q_a^{t+\Delta t} = Q_b^{t+\Delta t} + Q_s^{t+\Delta t} \tag{12}$$

$$H_a^{t+\Delta t} = H_b^{t+\Delta t} \tag{13}$$

where $H_{a,b}$ is the piezometric head along pipeline just before and after hybrid vessel (m) and $Q_{a,b}$ is the water flow in pipeline just before and after the hybrid vessel (m^3/s).

The rate at which the pump decelerates after losing power primarily depends on the rotational speed and inertia of the pump. The sum of the inertias, I_{tot}, of the rotating components of the pump is determined by [35]:

$$I_{tot} = I_{motor} + I_{pump\ impeller} = 118\left(\frac{P}{N}\right)^{1.48} + 1.5 \times 10^7\left(\frac{P}{N^3}\right)^{0.955} \tag{14}$$

where I = inertia (N-m^2); N = the speed of the pump rotation (rpm), with an average value of 1500 rpm for wind range of centrifugal pumps; and P = pump power (KW) given by the following:

$$P = \frac{\rho g Q H_{pump}}{3.6 \times 10^6 \eta} \tag{15}$$

where η = total efficiency with average value of 0.85, and Q = flow rate (m^3/h); H_{pump} = pump head at operating point (m). Pump operation in different zones is represented using the four-quadrant as follow:

$$N_s = \frac{N\sqrt{Q}}{{H_{pump}}^{3/4}} \tag{16}$$

where Q is pump flow at operating point (L/s). The pump and the surge vessel boundary conditions are numerically solved using the MOC [22,35], and the air volume and water level fluctuations inside the hybrid vessel are modelled as developed by [30,31].

2.3. *Hybrid Surge Vessel Optimum Sizing Approach*

Using the above equations and boundary conditions, the water hammer problem in pumping mains is solved numerically for: the expanded volume of the air $\forall_{air}^f$, the volume of the compression chamber $\forall_{cc}$, and eventually the total volume of the hybrid surge vessel $\forall_{tank}$. The maximum and minimum permitted pressure surges are identified and an iterative approach is used to optimize the solution within those bounds. In every iteration, the numerical model is solved using the Method of Characteristics for the transient pressure wave along the pipeline. In other words, given a series of model inputs, the results (volume of compression chamber, initial volume of air, and final tank volume) are uniquely determined. This deterministic case covers only a unique pipeline profile with related hydraulic characteristics.

However, to create enough data that describe the process of hybrid surge vessel operation during the protection cycle, a broad range of input pipeline and pump parameters with diverse random combinations would be required. Thus, in this study, a stochastic technique utilizing the Monte Carlo Simulations is used to propagate input parameter uncertainty via the numerical model in order to predict the associated volumes of the hybrid surge vessel. The framework of surge attenuation and control in a hybrid surge vessel using stochastic Method of Characteristics as shown in Figure 2:

1. A uniform probability distribution is allocated to input parameters for sampling within predetermined realistic ranges.
2. According to a predetermined number of Monte Carlo Simulations, random samples are taken from input parameters' distribution.
3. For each stochastic run, which is a different pipeline profile and hydraulic situation, combinations of random variables are inserted into the numerical model, which is then solved deterministically for the pressure change throughout the pipeline. In order to optimize the volume of the hybrid surge vessel, each deterministic run is repeated through numerous iterations.
4. The least-square linearization (LSL) method is used to determine the lowest influencing input parameters, which may be eliminated without affecting accuracy, using the result of the Method of Characteristic in the previous phase using the parameter coefficient of sensitivity, S_{V_i}:

$$S_{V_i} = \frac{100 \times w_i^2 \sigma_{\Delta v_i}^2}{\sum_{i=1}^{n} w_i^2 \sigma_{\Delta v_i}^2} \tag{17}$$

   where w_i is the coefficient of regression, $\Delta v_i = v_i - m_{vi}$, difference between parameter and mean, and $\sigma_{\delta_{v_i}}^2$ is the variance of Δv_i.

5. Then, an approximation relation can be generated using machine learning utilizing generated matrices with random parameters and the associated optimal vessel volume.

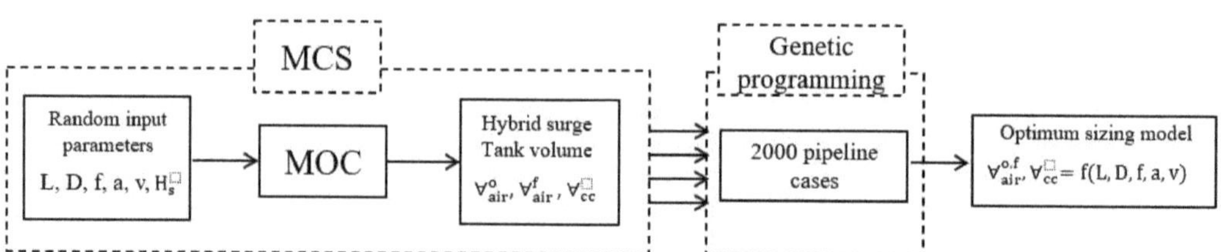

Figure 2. Stochastic Method of Characteristics/Monte Carlo Simulation framework for hybrid vessel sizing model development.

2.4. Sizing Predictive Model Development

The required volumes can be obtained in a simple relation as a function of the related input parameters generated by the Monte Carlo Simulation and the Least-square linearization, MCS/LSL. In many hydraulic engineering applications, genetic programming (GP) has outperformed other machine learning techniques such as artificial neural networks (ANN). It has shown its superiority in predicting complex relations between parameters using simple equations. The genetic programming code developed by [36] is utilized in this study. According to Table 1, the genetic programming developed model is evaluated based on a set of performance metrics.

Table 1. Performance indicators used for validation and testing of developed models.

Performance Indicator	Equation	Variables
Coefficient of determination	$R_i^2 = \left(\frac{\frac{1}{n}\sum_{j=1}^{n}(T_j-\overline{T})(P_{(ij)}-\overline{P})}{\sqrt{\sum_{j=1}^{n}(T_j-\overline{T})^2/n}\sqrt{\sum_{j=1}^{n}(P_{(ij)}-\overline{P})^2/n}} \right)^2$	
Root mean square error	$RMSE = \frac{\sum_{j=1}^{n}(T_i-P_i)^2}{n}$	
Coefficient of efficiency	$E_{sn} = 1 - \frac{\sum_{i=1}^{n}(T_i-P_i)^2}{\sum_{i=1}^{n}(T_i-\overline{T})^2}$	
Index of agreement	$D_{ag} = 1 - \frac{\sum_{i=1}^{n}(T_i-P_i)^2}{\sum_{i=1}^{n}(\lvert P_i-\overline{T}\rvert+\lvert T_i-\overline{T}\rvert)^2}$	$P_{(ij)}$ is the value predicted by the program, T_j is the target value, and n is the number of samples, $\overline{P} = 1/n\sum_{j=1}^{n}P_j$, $R_O^2, R_O'^2 = 1 - \sum_{i=1}^{n}P_i^2(1-k',k)^2/\sum_{i=1}^{n}(P_i-\overline{P})^2$
Gradients of the regression	$k = \sum_{i=1}^{n}(T_i \times P_i)/P_i^2$ or $k' = \sum_{i=1}^{n}(T_i \times P_i)/T_i^2$	
Slope of regression line	$m\prime = \left(R^2 - R_O^2\right)/R^2$ and $n\prime = \left(R^2 - R_O'^2\right)/R^2$	
Cross testing coefficient	$R_m = R^2 \times \left(1 - \sqrt{\lvert R^2 - R_O^2\rvert}\right)$	
Standard deviation errors	$S_e = \sqrt{1/(n-1)\sum_{j=1}^{n}(e_j-\overline{e})^2}$	$e_j = P_j - T_j$

3. Results and Discussions

3.1. Model Setup—Deterministic Run

As previously mentioned, a single deterministic run for the hybrid surge vessel's volume is an optimization process that involves repeatedly solving the water hammer equations until the ideal size is determined per predetermined pressure constraints. A schematic of a typical pumping problem with a flat profile and a downstream reservoir is shown in Figure 3. The steady state pump head H_{pump}, is the summation of the pipeline static head H_s, and the pipe friction loss head, H_f. The hybrid surge vessel is placed shortly after the pump. A hypothetical case is considered for a low-head profile and governing down surge, with the following data; a = 750 m/s, v = 1.5 m/s, L = 3000 m, D = 1.0 m, H_s = 5 m, f = 0.015, D_{con} = 0.75 m, and n =1.2. As discussed before, in low-head profiles, the negative pressure surge governs the hybrid surge vessel's optimization process is presented with H_{min} equal to −5.2 m following a sudden pump failure without surge protection.

The vessel connecting pipe is throttled be decreasing its diameter compared to the main pipe diameter with a recommended ratio of 70%. On the other hand, there would be no need to throttle the inflow to the vessel, which would not affect vessel size in limiting negative transients in low-head pipelines.

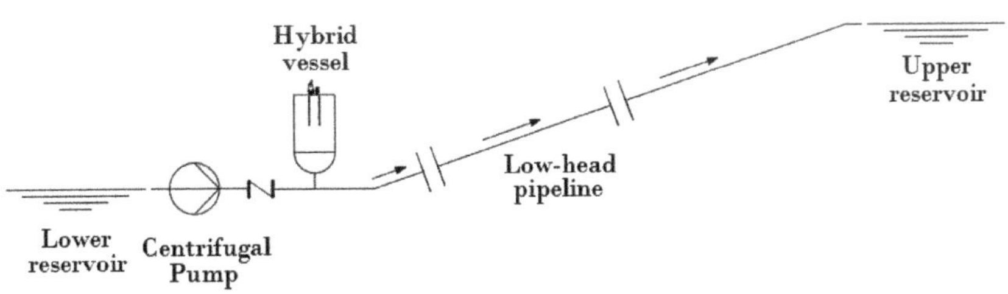

Figure 3. Schematic of pumping main problem with hybrid vessel for surge protection.

Sizing of the hybrid surge vessel starts with assuming a normal closed surge vessel without a ventilation tube, where the initial air volume $\forall^{o}_{air}$, and expanded air volume $\forall^{f}_{air}$, and the final tank volume $\forall_{tank}$ can be calculated using the model developed in [22]:

$$\forall^{o}_{air} = 0.11(D_{con}/D)^{-3} * v^{0.11} * D^{2.38} * (L/H_s)^{0.20} * f^{-1} \quad (18)$$

$$\forall^{f}_{air} = 1.2 * \forall^{o}_{air} * (L/H_s)^{0.08} \quad (19)$$

where $\forall^{o}_{air}$ = initial volume of air in hybrid vessel before transient event (m^3); $\forall^{f}_{air}$ = maximum expanded air volume in the hybrid vessel (m^3); D_{con} = diameter of hybrid vessel connecting pipe (m); L = pipeline length (m); H_s = static head (m); and v = steady flow velocity in pipelines (m/s).

3.2. Hybrid Surge Vessel Sizing

The hybrid surge vessel reduces volume compared to the normal surge vessel in cases of down surge transients. From the authors' practical experience and thousands of numerical simulations, the reduction in volume would be in the range of 10% to 50% depending on the specific case parameters. Therefore, an excellent initial guess for the volume of the hybrid surge vessel, $\forall_{tank}$, would be 70% of the volume calculated from Equation (19). Work in this paper shall follow [25] commercial ARRA-hybrid vessel size, which is limited to 35 m^3 per vessel. Larger vessels would be possible; however, they would be custom made and this would affect hybrid vessel choice from an economical point of view. Thus, the number of hybrid vessels, N_{tanks}, for a specific case shall be calculated from the following relation:

$$N_{tanks} = \begin{cases} 1 & \text{if } \forall_{tank} \leq 35\ m^3 \\ \forall_{tank}/2 & \text{if } 35\ m^3 < \forall_{tank} \leq 70\ m^3 \\ \forall_{tank}/3 & \text{if } 70\ m^3 < \forall_{tank} \leq 105\ m^3 \end{cases} \quad (20)$$

Afterwards, the hybrid tank diameter, D_{tank}, and height, H_{tank}, can be determined such that $D_{tank} = 0.0651\forall_{tank} + 1 \leq 3m$. The diameter of the dipping tube is calculated as $D_{dt} = 0.15D_{tank}$ with a minimum value of 0.20 m to avoid small unreasonable tube diameters in case of small vessels. On top of the central ventilation tube, air release and vacuum valves are installed with diameters of 0.050 m and 0.080 m, respectively. Similar to the standard surge vessel, the numerical solution of the water hammer equations with surge protection commences with the assumption of the initial air volume $\forall^{o}_{air}$. In the hybrid surge vessel, additional vital parameters to be set before a solution are the total tank volume $\forall_{tank}$ and the volume of the compression chamber $\forall_{cc}$. From authors' experience, $\forall^{o}_{air} \approx 0.2\forall_{tank}$ and $\forall_{cc} \approx 0.2 - 0.5\forall_{tank} \geq \forall^{o}_{air}$ can be used as initial guesses. The height of the dipping tube can be directly calculated from the compression chamber volume. Table 2 shows iterations used to reach the economical hybrid vessel volume that satisfies the maximum positive and negative pressure head limits: $H_{max} \not> 1.4H_{pump}$ and $H_{min} \not< -3m$, respectively. In this study, an optimum solution $\forall^{f}_{air}$ in each deterministic run was calculated in 6 to 12 iterations. If $\forall^{f}_{air} \geq \forall_{tank}$, then the tank would be emptied and air would enter the pipeline causing cavitation, which is not a desirable hydraulic condition. In this case, we need a bigger tank volume. If $\forall^{f}_{air} < \forall_{tank}$ and the pressure limits are satisfied, then tank volume can be decreased and optimized more, as shown in the table. This optimization and reduction in the size of the hybrid vessel would have the stop condition of $\forall_{tank} = 1.10\forall^{f}_{air}$ to ensure the tank is always filled with water. In this example run, the volume of the hybrid vessel is reduced from 35 m^3 with $\forall_{tank}/\forall^{f}_{air} = 1.48$ to an optimum value of 23 m^3 with $\forall_{tank}/\forall^{f}_{air} = 1.11$ through 10 iterations. For protection against negative pressure surges, water is pushed from the hybrid vessel with high rate to compensate for sub-atmospheric pressure in the pipeline and the atmospheric air enters the vessel when the water level in the vessel drops below the dipping tube bottom. Thus, the final expanded air volume, $\forall^{f}_{air} = 20.5m^3$, which is around 10 times the initial air volume inside the vessel, is $\forall^{o}_{air}$ =1.8 m^3. The atmospheric pressure compensated for most of this expanded volume through the ventilation tube.

Table 2. Iterations required for reaching optimal hybrid vessel volume.

Iteration Number	N_{tank}	$\forall_{tank}$ (m^3)	H_{tank}(m)	D_{tank}(m)	$\forall^{o}_{air}$ (m^3)	$\forall_{cc}$ (m^3)	$\forall^{f}_{air}$ (m^3)	$\forall_{tank}/\forall^{f}_{air}$	H_{max}/H_{pump}	H_{min} (m)
1	2	35	5.00	3.00	1.00	7.00	21.36	1.64	1.03	−5.21
2	2	35	5.00	3.00	1.50	7.00	21.76	1.60	1.05	−4.51
3	2	35	5.00	3.00	2.00	7.00	22.68	1.55	1.04	−3.70
4	2	35	5.00	3.00	2.50	7.00	23.51	1.48	1.04	−3.00
5	2	26	5.35	2.50	1.50	5.25	21.00	1.23	1.05	−3.34
6	2	26	5.35	2.50	2.00	5.25	21.72	1.20	1.04	−2.51
7	2	24	4.95	2.50	1.50	4.75	20.39	1.17	1.15	−3.59
8	2	24	4.95	2.50	1.75	4.75	20.74	1.15	1.14	−3.12
9	2	24	4.95	2.50	1.80	4.75	20.90	1.14	1.15	−2.91
10	2	23	4.75	2.50	1.80	4.62	20.50	1.11	1.21	−3.00

In the above case, two hybrid tanks were required to protect the pump main from down surge pressure, considering the size constraint set in Equation (20). When using multiple hybrid tanks for protection against down surge, their hydraulic operation must be identical so that each would supply the same volume of water to the pipeline and reach the same expanded air volume, $\forall^{f}_{air}$, in the vessel at the same time instance. This requires that the height of all tanks, H_{tank}, and the ventilation tube bottom level, Z_{dp}, are the same. Accordingly, the multiple tank operation shall be identical to having a single tank with a summation of their volumes. Table 3 shows the initial expanded air and tank volumes for two hypothetical cases in the case of using one big hybrid tank versus multiple tanks for protection. Case one; a = 750 m/s, v = 2 m/s, L = 4000 m, D = 0.90 m, H_s = 6.5 m, f = 0.015, D_{con} = 0.65 m, and n =1.2. Case two; a = 750 m/s, v = 2.25 m/s, L = 5000 m, D = 1.0 m, H_s = 10 m, f = 0.014, D_{con} = 0.85 m, and n =1.2. The tank height is considered 4.3 m constant for one big tank and multiple small tanks, and the ventilation tube bottom is constant for all tanks. It is observed that the final expanded air volume in all tanks is the same, and their summation is equal to the expanded air volume for one big tank, with a minor difference of 1%.

Table 3. Single versus multiple hybrid surge vessels.

Case	N_{tank}	D_{tank}(m)	H_{tank}(m)	$\forall_{tank}$(m^3)	$\forall^{o}_{air}$(m^3)	$\forall^{f}_{air}$(m^3)	$\sum\forall^{f}_{air}$(m^3)
1	1	3	4.3	60	7.2	53.96	53.96
	2	2.95		30	3.6	27.23	54.46
	3	2.45		20	2.5	18.46	55.38
2	1	* 5.2	4.3	90	18	87.24	87.24
	3	3		30	6	29.47	88.41
	4	2.6		22.5	4.5	21.59	86.36

Note: * Diameter is not feasible from manufacturing point of view, but shown as illustration.

3.3. Hybrid Surge Vessel Versus Standard Surge Vessel

To study the hydraulic performance of the hybrid tank versus that of the surge tank, the previous down surge case is considered with the protection of each tank. The needed final surge vessel volume was 102 m^3, while the hybrid vessel volume was smaller by 30%, with $\forall_{tank}$ = 70 m^3. Figure 4 shows the changes in air volume inside the normal and hybrid surge vessels during a protection cycle following a down surge caused by pump failure. The hybrid tank's economic size is attributed mainly to the much smaller initial air volume required, which is 4 m^3, compared to the surge tank, which requires an initial air volume of 50 m^3. Accordingly, the expanded air volume reached 40 m^3 in the hybrid vessel and a larger value of 91 m^3 in the surge vessel. In the surge vessel, compensation for low pressure in the pipeline depends mainly on the initial air volume inside the vessel before the surge event, which would push water from the tank to the pipeline with the required volume. The hybrid tank works by the same principle of pushing water to the pipeline, but the ventilation tube would allow the atmospheric pressure to support releasing water from the tank when the water level drops below the compression chamber level. It is to be noted that the efficiency of the hybrid vessel is shown in the expanded air volume that reached 10 times the initial air volume, while it reached only 1.8 times in the surge vessel. However, the amount of water supplied to the pipeline, Q_s, during the first peak of the negative pressure wave for both tanks are the same. It can be calculated from Figure 4 as the difference between the final expanded air volume and the initial air volume, which has an average value of 38 m^3. This is also shown in Figure 5, which demonstrates the water flow from the vessel to the pipeline during the transient protection case. Fluctuations of water flow from and to the pipeline are lower in the case of a hybrid vessel than the surge vessel, as shown in Figure 5, and as supported by vessel air volume fluctuations in Figure 4.

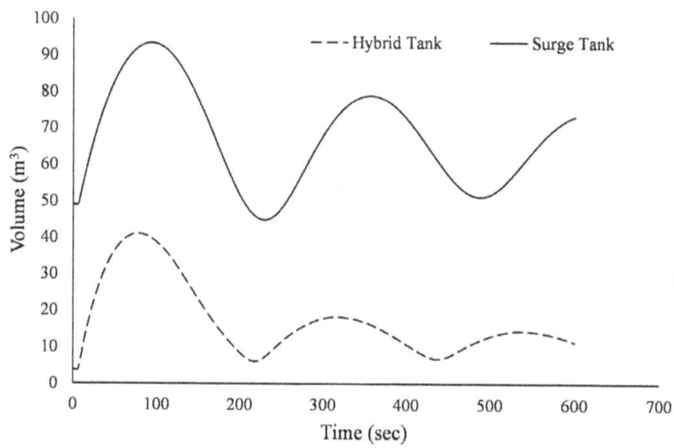

Figure 4. Changes in air volume during down surge protection inside surge vessel and hybrid surge vessel for the same pump failure case.

Water level fluctuations inside the hybrid and surge vessels during surge protection are shown in Figure 6. Both tanks are designed to have a 10–15% margin of volume safety so that water would not be emptied during a surge event. Therefore, during the first peak of the negative wave, the maximum drop of the water in both tanks is close for both tanks at an average level of +1.5 m, which is higher than the bottom level of tanks by +0.5 m. Considering the initial water level in both tanks, the relative water level fluctuations are similar. During the complete negative transient event, the hybrid vessel has been working as a vented vessel, where the water level in the hybrid vessel was below the compression chamber's lower level of +4.90 m. The compression chamber contained water inside the vessel at high pressure instances of the reflecting negative transient wave.

Figure 7 shows the air pressure variations inside hybrid and surge vessels, calculated in m H_2O. Initially, both tanks' water level and air pressure above water is maintained by the steady-state hydraulic grade line of the pump main. Thus, the air pressure inside the two tanks is equal, with a slight difference of 2 m accounting for the difference in water level between the two tanks. After a pump fails and down surge waves are created, both tanks release water into the pipeline, which lowers the air pressure inside the tanks.

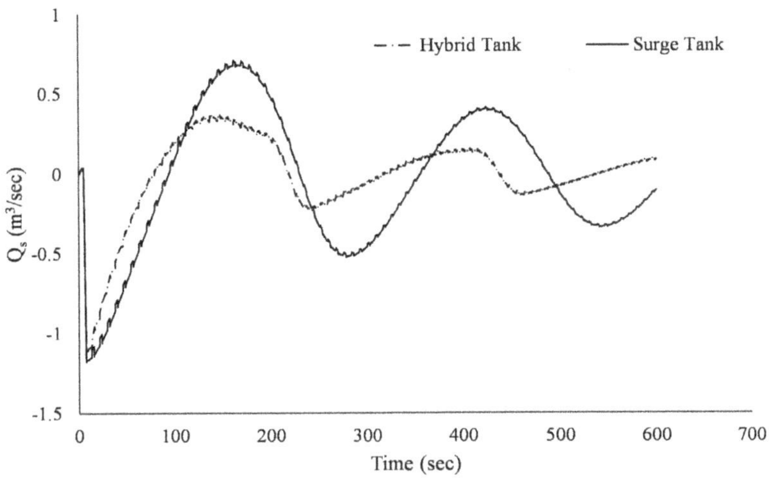

Figure 5. Changes in vessel water flow during surge protection for surge and hybrid vessels for the same pump failure case.

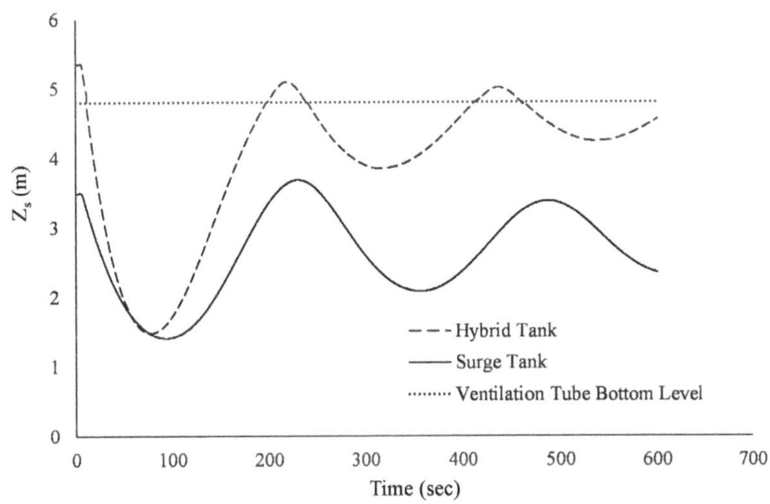

Figure 6. Changes in water level inside surge and hybrid vessel during surge protection for the same pump failure case.

The drop is faster in the case of the hybrid vessel and reaches a low air pressure value of 6 m H_2O following operation of the vented tube, while this drop is gradual in the case of the surge vessel with a longer time span. The fast drop in air pressure in the hybrid vessel is due to the smaller initial air volume [31]. It is observed that the air pressure inside the hybrid vessel vacuum valve maintained an atmospheric value of 10 m H_2O during the whole down surge following the initial drop, except when water level inside the hybrid vessel passed the ventilation tube bottom, as seen in Figure 6. The compression chamber is closed by water below it, the air is pressurized above atmospheric value, and the air release valve in the ventilation tube works to maintain air pressure in the tube as the air pressure in the compression chamber.

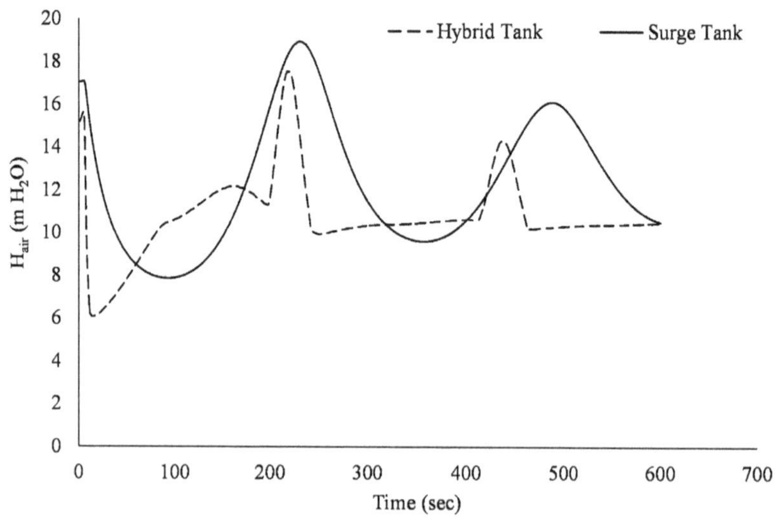

Figure 7. Variations in air pressure inside surge and hybrid vessel during surge protection for the same pump failure case.

As discussed, the ventilation tube allows the hybrid surge vessel to effectively compensate for negative pressure in low-head pipelines using minimum initial air volume compared to a normal surge vessel. Thus, lower total vessel volume would be needed as shown in a previous case, in which the hybrid vessel's initial air volume was 4 m^3 and the surge vessel initial air volume was 50 m^3. However, in some low-head pipelines, the initial air volume required for down surge protection in the normal surge vessel would be close to that required in the hybrid surge vessel. In these cases, the expanded air volumes for the two vessels would be similar, implying similar tank volumes and, in some cases, larger hybrid vessel volumes. This can be observed in Figure 8, which shows the variation in air volume inside a normal and hybrid surge vessel. The pump failure case parameters are: a = 1000 m/s, v = 2.5 m/s, L = 6000 m, D = 1.0 m, H_s = 6 m, f = 0.015, D_{con} = 0.85 m, and n =1.2. The initial air volumes for the surge and hybrid vessels were close; 30 m^3 and 24 m^3, respectively. Both vessels compensated for pressure loss in the pipeline, as shown in Figure 9, and kept H_{min} above −3 m. The expanded air volume following the first wave of the down surge was higher in the hybrid vessel by 7% compared to the surge vessel, with values of 140 m^3 and 130 m^3, respectively. Many trials for different ranges of hydraulic and pipe parameters were attempted; results showed that the hybrid vessel volume was larger than the surge vessel volume by about 50% for some cases. Thus, despite being a low-head pipeline with a down surge governing transient, the hybrid vessel would not be economical for design engineers.

Figure 10 shows the expanded air volume in cases using the normal and the hybrid surge vessels on the above test case, calculated for various pipe slope values H_s/L. It is observed from the figure that the hybrid vessel is an economical choice compared to the surge vessel for pipe slope $H_s/L \leq 0.0025$. For other values of pipe slope $H_s/L > 0.0025$, both vessels' final air volume starts to close until reaching the inflection point at $H_s/L = 0.0035$, where the hybrid vessel volume becomes more than the surge vessel. Analyses showed that a hybrid vessel would be most economical to use on low-head pipelines with slopes $H_s/L \leq 0.001$–0.003. This pipe slope range was found to be highly dependent on pipe diameter and flow velocity.

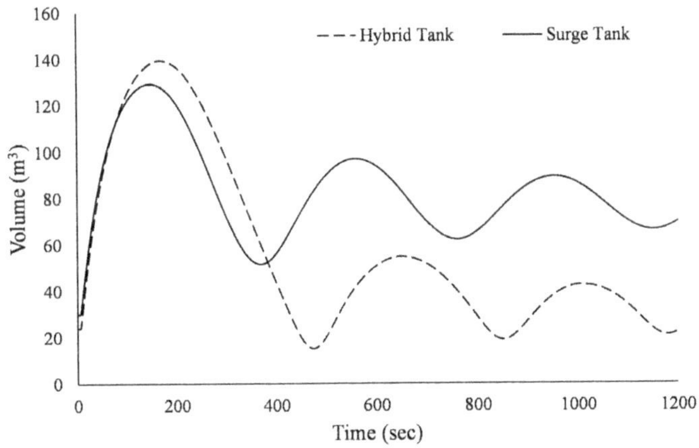

Figure 8. Changes in air volume inside normal and hybrid surge vessels for the same pump failure case.

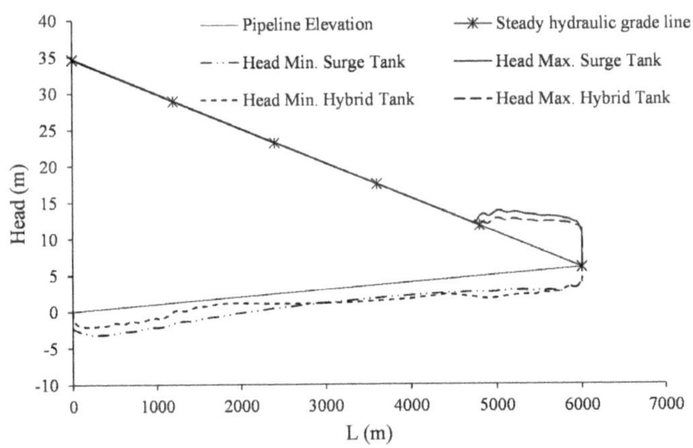

Figure 9. Maximum and minimum pressure heads with protection using normal and hybrid surge tanks.

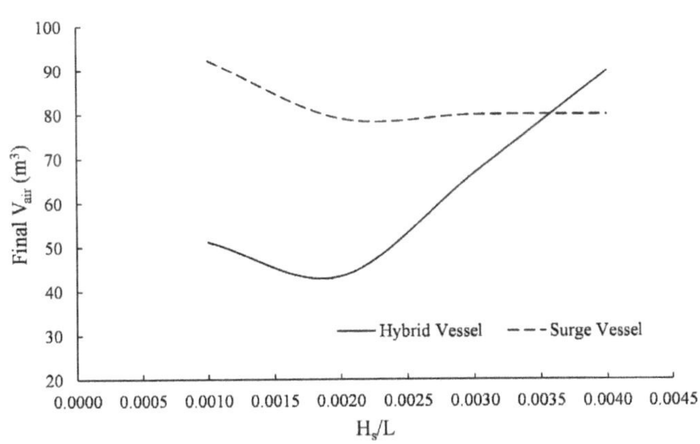

Figure 10. Expanded air volume for surge vessel and hybrid vessel versus pipeline slope.

3.4. Model Setup—Stochastic Runs

The stochastic Method of Characteristics numerical analysis is presented in this section. Table 4 shows the pumping main cases considered in this work and their range of input parameters. The steady-state velocity is used thereafter instead of discharge to describe flow in system. This is similar to previous research on surge vessels, in which surge vessel size is presented in terms of the steady velocity, e.g., refs. [13–15]. The steady-state velocity varies from 0.5 m/s to 2.5 m/s, and the pressure wave celerity from 250 to 1400 m/s. Various types of pipe material, from plastic to concrete, are modeled with Darcy–Weisbach friction coefficient from 0.015 to 0.3, length of pipe main from 2500 m to 15,000 m, and diameter from 0.25 m to 2.0 m. The specific speed of pumps is calculated according to Equation (16). Radial-flow and mixed-flow pumps are considered with specific speeds of 600 to 8000 (US units). The initial discharge is set as the pump operating point on the pump and system curves. Various discharge values correspond to a random choice of pumps.

Table 4. Range of input parameters used in the Monte Carlo simulations.

Parameter	Range	
	Lower Bound	**Upper Bound**
f	0.015	0.030
L (m)	2500	15,000
v (m/s)	0.5	2.5
D (m)	0.25	2.0
H_s (m)	5	40
a (m/s)	250	1400

Various input parameters for the hydraulic problem of water hammer and hybrid vessel protection are modelled, considering equal probabilities of inputs. Thus, the uniform probability distribution density function is evaluated and bounded by the range of parameters shown in Table 4. A surge index has been developed to differentiate between upsurge and down surge cases based on work on surge vessels [22]. The index has been extended and modified in this work according to the hybrid vessel with dipping tube operation mechanism to define only down surge cases, I_{DS}, as follow:

$$I_{DS} = 7.89(D \times H_s/L)^{0.2} \times 1.44^{-v} \leq 1.6 \tag{21}$$

For each of the input parameters, 5000 cases are randomly generated, such that all generated cases would comply with the down surge index set, I_{DS}. When each random case scored less than 1.6 in the above index, a down surge condition would govern the transient problem. The hybrid vessel would probably be more economical to use than normal surge vessel. Many random combinations of input parameters are prepared. Each random realization is used as input to the numerical model to solve as a single deterministic Method of Characteristics run; 5000 deterministic simulations using a new random combination of input parameters were considered. If the final volume of the hybrid tank is less than that initially calculated in case of a surge tank then the run is considered; otherwise, the run is neglected. Out of the initial 5000 cases, only 2500 cases were considered. The accuracy of the stochastic Monte Carlo simulation solution depends largely on the realization number considered. The accuracy of the Monte Carlo simulation is probed using the variance of the output, which is the initial and expanded air volumes in our case. It is observed that the variance started to reach a steady value after 1800 Monte Carlo realizations. Therefore, the 2000 random cases would be accurate enough for utilization in the stochastic analyses. From Figure 11, the calculated initial, expanded, and compression chamber volumes is best fitted with the generalized extreme distribution. Most of the optimized vessels had an initial air volume ≤ 20 m^3, and the expanded air volume ≤ 150 m^3, and the compression chamber volume ≤ 65 m^3.

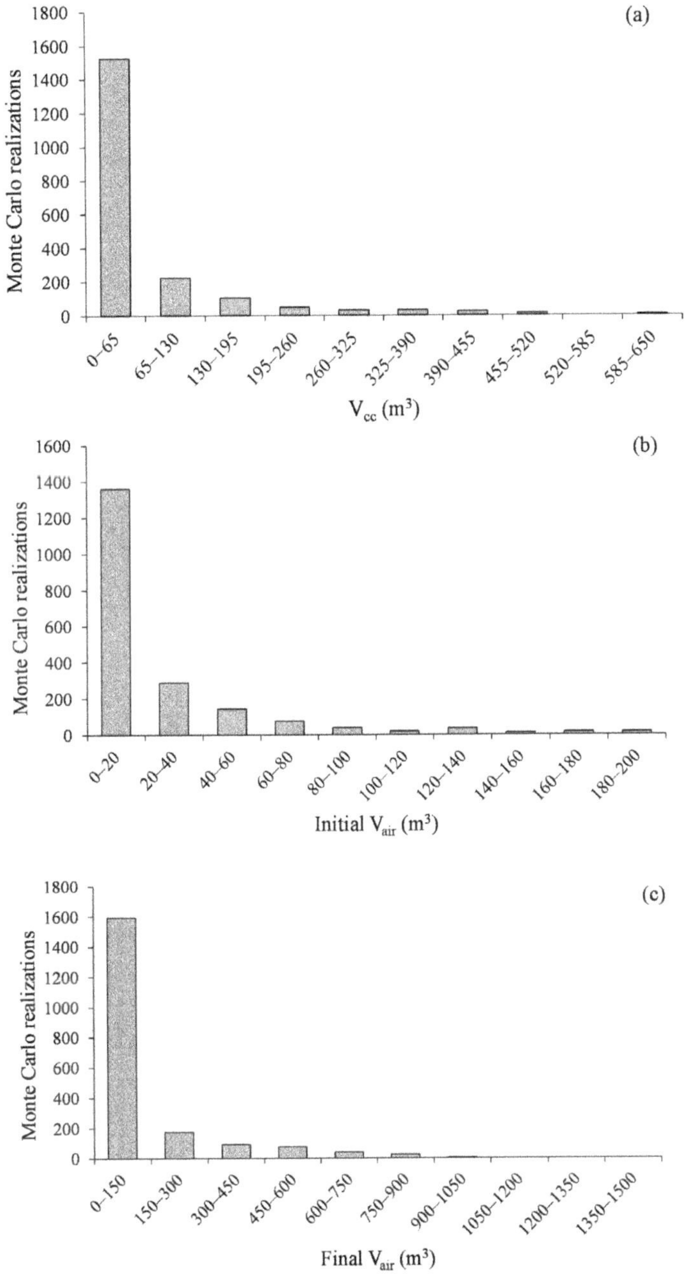

Figure 11. Probability distributions of (**a**) the calculated compression chamber; (**b**) initial; and (**c**) expanded air volumes.

3.5. Parameters Selection

From the authors' practical experience and previous literature, the compression chamber volume and initial/expanded air volumes for the hybrid vessel can be a function of the following flow and pipe parameters:

$$\forall = F(v, a, D, L, f, H_s, H_{max}, H_{min}, \forall_{cc}) \tag{22}$$

where v = flow velocity (m/s); a = wave speed (m/s); D = pipeline diameter (m); L = pipeline length (m); H_s = static head (m); $H_{max,\ min}$ = pressure limits (m) and $\forall_{cc}$ = compression chamber volume (m^3). The maximum allowable pipe pressure, $H_{max,}$ has been set to 1.4 H_{pump}, while the minimum pressure, $H_{min,}$ has been set to a value of -3 m. Other parameters controlling the air volume and, thus, the hybrid vessel volume have varying effects on the volume; however, their contribution to the hybrid vessel initial and expanded air volumes has not been discussed before. Using the least-square realization method, various flow and pipe parameters are ranked according to their contribution to the final tank volume. Table 5 shows the least-square realization regression hybrid vessel sensitivity results, S_{V_i} for various input parameters, as presented in Equation (22). It is found that the compression chamber volume, $\forall_{cc}$, is largely sensitive to the pipeline's diameter with the contribution of 94% followed by the system static head, H_s with 2% and the pipe friction coefficient with 3%, other parameters; L, v, and a had negligible impact on the chamber volume. On the other hand, the compression chamber volume contributed to the initial air volume, $\forall^o_{air}$, as an input parameter with more than 95%; with all other input parameters contributing 5%, the highest impact was for the pipe diameter with 2.8%, followed by the wave speed with 1.7%, and the static head with 0.4%. The friction coefficient and flow velocity had zero contribution to the initial air volume. Results in Table 5 show that the wave velocity had the least contribution to the expanded air volume. Other parameters had a larger contribution; the pipe diameter was the highest with 89%, followed by the velocity with 3%, pipe length with 2.7%, and pipe friction with 3.7%. Calculation showed that the hybrid tank volume is mostly influenced by the pipe diameter as the major contributing parameter, followed by the line static head.

Table 5. Least-square regression-vessel volume sensitivity results, S_{V_i} in percentage.

Volume (m^3)	f	L (m)	v (m/s)	D (m)	H_s (m)	a (m/s)	$\forall_{cc}$ (m^3)
$\forall_{cc}$	3.38	0.001	0.05	94.23	2.32	0.01	--
$\forall^o_{air}$	0.001	0.20	0	2.80	0.38	1.75	95.11
$\forall^f_{air}$	3.7	2.71	3.43	89.39	0.66	0.066	--

3.6. Developed Genetic Programming Models

Using the most influential input parameters calculated in Table 5, the machine learning-based models were developed for the compression chamber volume, initial air volume, and the final expanded air volume. Correlation analysis was conducted between all input predictive variables, and no interdependency was found amongst them. Thus, any problem that could arise in the model development due to dependency between input parameters is omitted. The random database was divided into training and testing sets with consistent statistical measures of both sets. The 2000 cases are divided randomly into a set used for model training (1500 cases) and another set for testing the developed model (500 cases). The algorithm defined the parameter setting with a mutation rate of 0.06, a chromosome number of 25, and a transposition rate of 1. The developed Genetic Programming models for the compression chamber volume, the initial air volume, and the expanded air volume can be written as follows:

$$\forall_{cc} = D^3 H_s^{1.04} f^{-0.33} \tag{23}$$

$$\forall^o_{air} = 5.127 D^{-0.333} H_s^{-0.111} a^{-0.33} \forall_{cc} \tag{24}$$

$$\forall^f_{air} = D^{2.667} H_s^{-0.4} v^{1.2} L^{0.5} f^{-0.05} \tag{25}$$

where $\forall_{cc}$ is the compression chamber volume (m^3), $\forall^o_{air}$ is the initial gas volume (m^3), $\forall^f_{air}$ is the expanded air volume (m^3), v is the steady state velocity (m/s), D is the pipeline diameter (m), L is the pipeline length (m), H_s is the static head (m), and f is the Darcy–Weisbach friction coefficient. The compression chamber volume is an influential predictive input for the initial air volume Genetic Programming model only, as shown in Equation (23).

Table 6 shows the general statistical performance of the developed hybrid vessel compression chamber, initial air, and expanded air volumes. The developed models were scored in R^2, and an average of 0.92 for training and 0.91 for testing datasets. The error measures, root mean square error, RMSE, and root absolute error, RAE, showed close figures of training and testing datasets for compression chamber and expanded air volumes. The root mean square error for $\forall_{cc}$ is 33 and 44 for the training and testing sets, respectively. The root absolute error is 0.25 and 0.33, respectively, for training and testing sets. The root mean square error for $\forall^f_{air}$ is 32 and 40, respectively, for training and testing sets, while the root absolute error is 0.14 for both the training and testing sets. the error measures showed lower values for the initial air volume. The root mean square error for $\forall^o_{air}$ is 8 and 10, respectively, for training and testing sets, while the root absolute error is 0.19 and 0.21, respectively, for training and testing sets. On the other hand, the indices E_{sn} and D_{ag} showed satisfying values with averages of 0.92 and 0.89 for the training and testing subset, respectively. Figure 12 shows the Q-Q plot of the Method of Characteristics numerical calculated volumes; $\forall_{cc}$, $\forall^o_{air}$ and $\forall^f_{air}$ versus those predicted by the developed genetic programming models using Equations (23)–(25). There was no substantial over- or under-prediction, and the created models demonstrated extremely good agreement with numerical estimated volumes with minor mistakes. The developed model for $\forall^o_{air}$ showed a slight under-predict volume with about 4%.

Table 6. Statistical performance of developed Genetic Programming prediction models.

Model	Data Partitioning	R^2	RMSE	RAE	E_{sn}	D_{ag}
[13]	NA	0.25	98	1.82	−6.48	−0.06
[15]	NA	0.17	339	4.92	−89.35	−0.67
GP-$\forall_{cc}$	Train	0.88	33	0.258	0.88	0.94
	Test	0.81	44	0.334	0.78	0.89
GP-$\forall^o_{air}$	Train	0.94	8	0.19	0.94	0.97
	Test	0.94	10	0.21	0.93	0.96
GP-$\forall^f_{air}$	Train	0.96	32	0.149	0.95	0.98
	Test	0.97	40	0.144	0.97	0.98

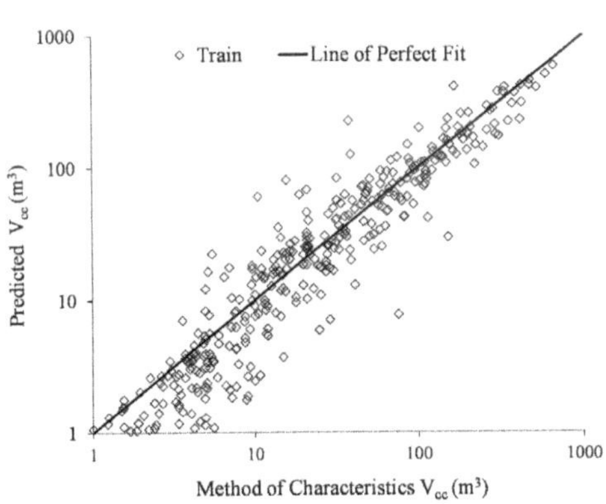

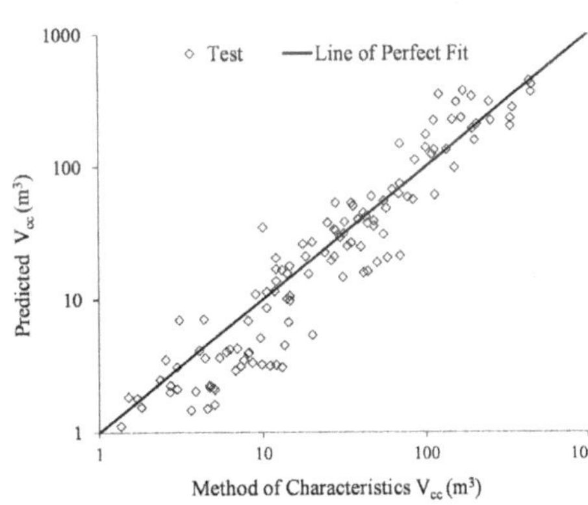

Figure 12. *Cont.*

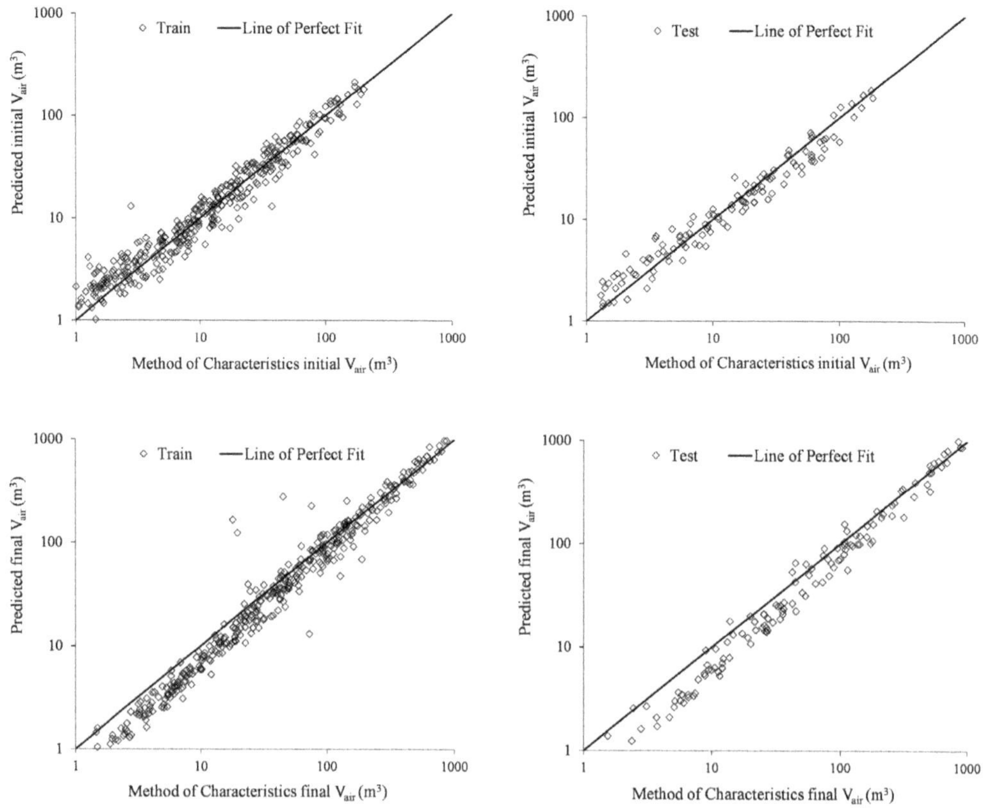

Figure 12. Method of Characteristic optimized $\forall_{cc}$, $\forall^{o}_{air}$ and $\forall^{f}_{air}$ versus that predicted by the genetic programming (GP) model.

3.7. Comparison with Classical Design Charts

Up to the authors' knowledge, there is no current research in literature for hybrid vessels. All existing research discusses the surge vessel performance and the initial air volume in these vessels. Two classical charts have been developed by [13,15] for calculating the surge vessel initial air volumes for rising pumping mains. They have been developed for normal surge vessels to protect rising pump mains from pump failures that are accounted for by a sudden valve closure and using the rigid column theory. They are regarded as traditional benchmark pillar models for the initial vessel air volume, and thus, it shall be used in this work for comparison with the developed model on a hybrid model, though it would be expected that they would oversize vessel volume. While it would be unfair to utilize the classical charts for comparison with the developed genetic programming models, the classical charts would provide rough benchmarking. Table 7 shows the range of parameters for the classical charts and the developed genetic programming models in this study, including the vessel initial air volume using classical non-dimensional parameters. Despite having charts developed without considering practical limits for their input parameters, the ranges of the non-dimensional parameters are very close.

Table 7. Range of parameters of classical design charts and the developed genetic programming models.

Parameter	[13]	GP-$\forall^{o}_{air}$	Parameter	[15]	GP-$\forall^{o}_{air}$
$S' = \frac{\forall^{o}_{air} a}{nALv}$	0 to 160	0 to 18	$S' = \frac{\forall^{o}_{air} g(H_s + H_{Bar})}{ALv^2}$	0.5 to 8	0.02 to 4.2
$\frac{H_{max} - H_s}{av/g}$	0.1 to 1	0.03 to 2.6	$\frac{H_{max} + H_{Bar}}{H_s + H_{Bar}}$	1 to 2.2	1.2 to 17.8
$\frac{H_{min} - H_s}{av/g}$	−0.8 to −0.1	−0.74 to −0.03	$\frac{H_{min} + H_{Bar}}{H_s + H_{Bar}}$	0.2 to 0.7	0.15 to 0.48

Both classical charts yielded an extremely low R^2 value, as shown in Table 6. The chart developed by [15] had the lowest R^2 value of 0.17, and 0.25 for [13] chart. Thus, the classical charts can only represent 20% of the data. The other statistical error values showed higher values when compared to the developed genetic programming models. The root mean square error for these classical charts showed values ranging from 98 to 339, which are 3 to 10 times higher than those of the genetic programming models. These errors are confirmed in Figure 13, which shows the scatter between the Method of Characteristics solution and the classical charts predictions. It is observed that the solutions by the classical charts show a spread of values around the line of fit with high over-prediction for initial vessel volume for the [15] chart. The chart in [13] was based on the rigid column theory, while [15]'s chart was based on the incompressible flow equations. An overestimation for the rigid column and incompressible flow theories in the initial vessel air volume computation were reported in [15]. Moreover, results in Table 6 showed that E_{sn} and D_{ag} indices for the classical charts yielded −89 and 0.5 respectively, indicating the charts' failure to predict the initial volume. Lower error values and excellent value of E_{sn} and D_{ag} indices were calculated using the developed models; these can predict more economical vessel sizes.

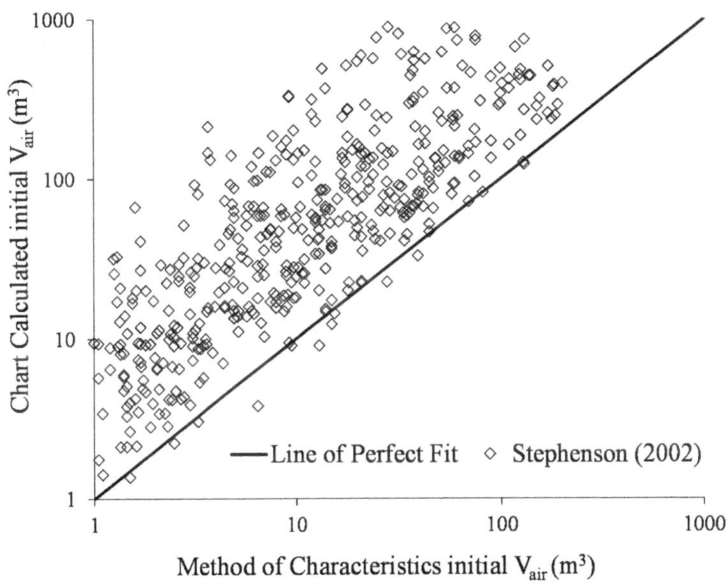

Figure 13. *Cont.*

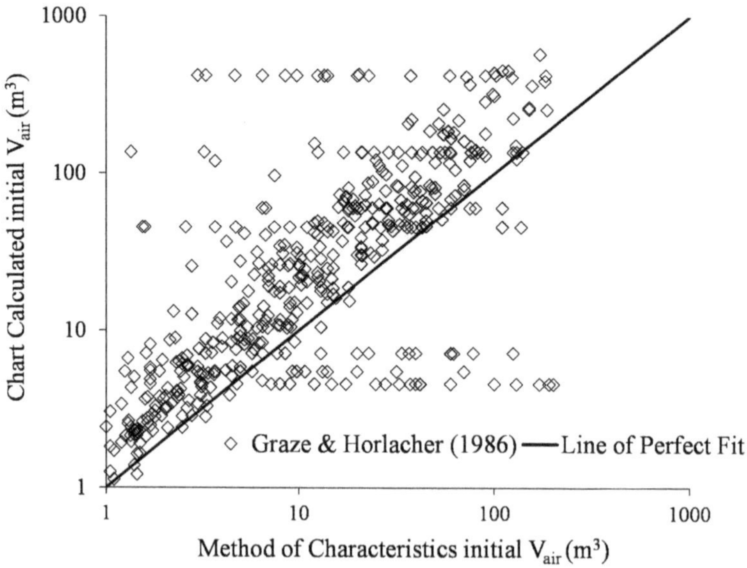

Figure 13. Method of Characteristics, MOC-optimized. $\forall^{o}_{air}$ versus that predicted by the classical design charts of [13,15].

3.8. Testing and Validation of the Developed Models

As shown in Table 8, model testing is applied to the test set. The slope of regression line between the predicted versus the Method of Characteristics calculated volumes were near 1 and in the suggested range of 0.85 to 1.15. In addition, models scored favorable values for the regression line's coefficient of determination, which had average values of −0.05. As a result, when tested against test datasets, the created model's performance can be said to be good and acceptable.

Table 8. Testing statistical measures for genetic programming prediction models (based on test dataset).

Model	R (R > 0.8)	K (0.85 < K < 1.15)	K′ (0.85 < K′ < 1.15)	m′ (m′ < 0.1)	n′ (n′ < 0.1)
GP-$\forall_{cc}$	0.96	1.01	0.97	−0.09	−0.09
GP-$\forall^{o}_{air}$	0.98	1.02	0.97	−0.03	−0.03
GP-$\forall^{f}_{air}$	0.99	1.04	0.95	−0.01	0.00

Table 9 displays the differences in prediction uncertainty between the created models and traditional charts. The table shows the average prediction errors of the different models, the width of the uncertainty band, and the error within the 95% confidence interval. It is based on the test data subsets. The developed models for $\forall_{cc}$, $\forall^{o}_{air}$, and $\forall^{f}_{air}$ showed a mean prediction error less than that of classical charts, with average values of 0.05. The classical charts had higher mean prediction errors, three and six orders of magnitude larger than the developed model for the initial air volume.

The genetic programming models prediction errors had a deviation of 1/2 to 1/4 that of the available classical charts, with values of 0.10 and 0.45, respectively. Since the classical charts were developed for the surge vessels on high-head pumping mains, they are expected to overestimate the initial air volume. This is confirmed by the mean prediction error positive signs in predictions of the [13,15] charts. The uncertainty with for the developed models are close, with values of +/−0.47, +/−0.26, and +/−0.25 for $\forall_{cc}$, $\forall^{o}_{air}$, and $\forall^{f}_{air}$, respectively.

Table 9. Uncertainty estimates for genetic programming prediction models (based on test dataset).

Model	Mean Prediction Error	Deviation of Prediction Error	Width of Uncertainty Band	95% Prediction Error Interval
[13]-$\forall^{o}_{air}$	+0.31	0.51	±1.01	+0.01 to +51
[15]-$\forall^{o}_{air}$	+0.66	0.40	±0.78	+0.01 to +7.88
GP-$\forall_{cc}$	−0.08	0.24	±0.47	+0.14 to +10.59
GP-$\forall^{o}_{air}$	0	0.13	±0.26	+0.30 to +3.30
GP-$\forall^{f}_{air}$	−0.12	0.13	±0.25	+0.43 to +4.14

3.9. Parametric Analysis of the Developed MODELS

Figure 14 shows the variation of the initial vessel volume versus pipeline diameter, static head, and the friction head as calculated by the genetic programming models, [22], and classical design charts of [13,15]. It is observed that the initial air volume calculated by the genetic programming model is the lowest amongst all other models since the genetic programming model is developed for the hybrid vessel, which would have a smaller volume than surge vessel in low-head pipelines. Moreover, the initial air volume increased with the increase in pipeline diameter for all models. The increase in the vessel volume is the smallest for $D \leq 0.60$ m, while it is more noticeable and with an increasing gradient for $D > 0.60$ m. The largest gradient of increase is observed for [13,37]'s chart, and the lowest is observed for the genetic programming model. When the pipe diameter is larger and the pump fails, a larger sub-atmospheric region is created. This would require more water to be pushed from the vessel to account for the low-pressure region. It was confirmed by [15] that the volume of the liquid zone formed in pipe with sub-atmospheric pressure.

According to [15,38], the pipe diameter and the volume of the liquid zone formed by the pump failure with sub-atmospheric pressure are directly proportional. For previous models, the initial air volume decreased with the increase in the static head with a near constant gradient for all models with the highest initial air volume, -due to [13] model, of around 150 m^3 compared to 50 m^3 for [15,22]. In contrast to earlier design models, the genetic programming model exhibited an opposing pattern in the change of the initial air volume with the pipe static head. The chances that the down surge wave following pump failure will fall below the pipe hydraulic grade is reduced as the static head of the pipeline rises. Thus, less initial air volume would be needed to compensate for the sub-atmospheric pressure in the pipeline in case of a surge vessel (as shown for previous models) and more initial air volume would be needed to act as a cushion inside the vessel against reflected upsurges (as shown in genetic programming model).

For the initial air volume variations in a vessel with the pipe friction head, as shown in Figure 14, a similar trend is observed for the genetic programming model and previous models, except for [13]'s chart, with a slight decrease in initial air volume when friction head increased. The trend of initial air volume calculated by the genetic programming model was close to that calculated by [15,22]. Relative to [13]'s model, all of these demonstrate a lower dependence on the initial air volume on the friction head. When the friction head increased by 40% (from 18 m to 25 m), the initial air volume decreased by an average of 5% for all models; this is compared to 50% for [13]'s model. The work of [37], who stated that an increase in pipe friction head would minimize positive transients and make it highly probable that the upsurge would never reach the original hydraulic grade, illustrates this declining trend.

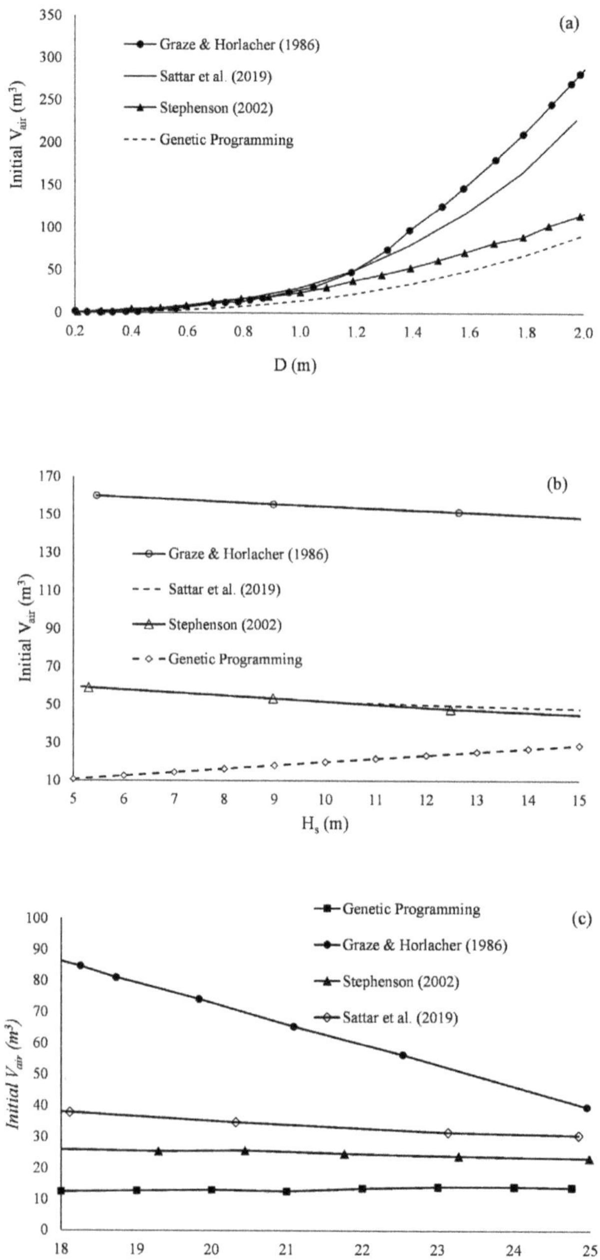

Figure 14. Method of Characteristics optimized initial air volume, $\forall^{o}_{air}$ versus: (**a**) D; (**b**) Hs; and (**c**) pipe friction losses as predicted by the genetic programming model and the classical design charts of [13,15] and GEP design model of [22].

3.10. Genetic Programming Models' Application Range

The developed genetic programming model can be used for providing hybrid vessel volume on linear varying low-head pipeline profiles with no intermediate high points and where down surge is the governing transient event. The applicability of the developed model has not been tested outside the range specified in Table 4. As discussed, the hybrid vessel might not be economical for some down surge cases. Thus, based on 2500 random cases, a genetic programming classification index,I_{Dt} has been developed (with R^2 of 0.70) to help guide design engineers to down surge cases where hybrid vessels would be most economical, as shown in Figure 15 and as follows:

$$I_{Dt} = D^{0.5}L^{0.25}H_s^{1.5}v^{-1}f^{1.5} \leq 0.45 \quad (26)$$

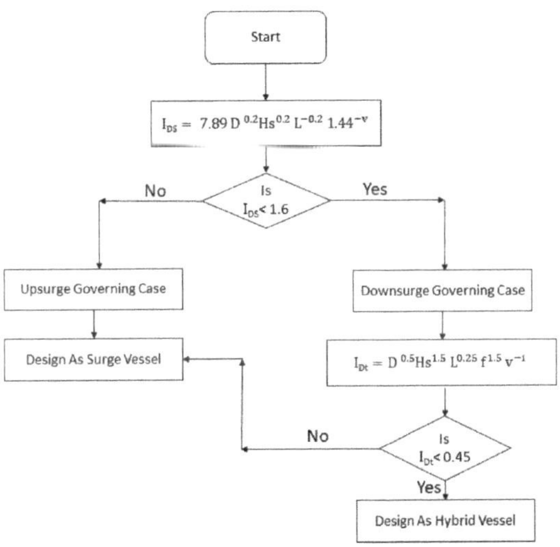

Figure 15. Hybrid Vessel selection based on developed down surge indices.

The proposed model has been built using radial-flow and mixed-flow centrifugal pump categories with the average rotational speed of 1500 rpm and average efficiency of 0.85. Using the genetic programming model would imply that no other surge protection device is installed on the pipeline, such as surge relief valves or vacuum breakers. Installation of these devices would benefit the system and interact with the surge vessel, which may further reduce the required size of the vessel. The genetic programming model developed using −3 m as set for the minimum head would allow backflow into pipelines during pump trip events at locations of existing leaks and thus might be a concern where water quality is a priority.

4. Conclusions

Surge vessels are known to be effective means of controlling pressure transients following pump failures. For low-head pipelines, the pump trip would create a sub-atmospheric pressure liquid region, which requires a large amount of water to flow from the vessel to compensate. Using a surge vessel would imply a high initial air volume and thus a high tank volume. However, the hybrid vessel can be utilized to effectively and efficiently compensate for the negative surge using the atmospheric pressure and lower initial air volume compared to the surge vessel. This research presented a hybrid vessel for protection against negative transients in low-head pipelines and developed a novel and simple approximate model for the estimation of the volumes of the compression chamber, initial air, and expanded air in the hybrid vessel. The developed model showed very good statistical performance, with a high coefficient of determination of 0.90 and a low mean prediction error of 0.1. The model clearly shows the relation between various hydraulic and pipe parameters with pipe diameter and static head as the most influencing parameters on compression chamber and air volumes. In addition, the developed model presented classification criteria for low-head pipelines on which the hybrid vessel would be most economical. The developed machine learning model with equations and indices can be utilized as a general framework used for the selection of hybrid surge vessels and their sizing. It can be associated with commercial software, e.g., Bentley HAMMER [34], in the design phase.

Author Contributions: Conceptualization, A.S.E.; Data curation, M.E.; Formal analysis, A.M.A.S. and A.N.G.; Investigation, A.M.A.S. and M.E.; Methodology, A.M.A.S.; Project administration, B.G.; Resources, M.E. and B.G.; Software, A.M.A.S. and A.N.G.; Supervision, B.G.; Validation, A.M.A.S. and A.N.G.; Visualization, A.M.A.S. and A.S.E.; Writing—original draft, A.M.A.S.; Writing—review and editing, B.G. All authors have read and agreed to the published version of the manuscript.

Funding: This research was funded by the Natural Sciences and Engineering Research Council of Canada (NSERC), Discovery Grant #400675.

Data Availability Statement: The data presented in this study are available on request from the corresponding author. The data are not publicly available due to having commercial real cases.

Conflicts of Interest: The authors declare no conflict of interest.

References

1. Laiq, S.J. Analyzing Transient Behavior and Cost Effectiveness of a TSE Transmission Line Considering Different Pipeline Materials. In *Pipelines 2014: From Underground to the Forefront of Innovation and Sustainability*; ASCE Library: Reston, VA, USA, 2014; pp. 2201–2210. [CrossRef]
2. Triki, A.; Chaker, M.A. Compound technique-based inline design strategy for water-hammer control in steel pressurized-piping systems. *Int. J. Press. Vessel. Pip.* **2019**, *169*, 188–203. [CrossRef]
3. Chaker, M.A.; Triki, A. Investigating the branching redesign strategy for surge control in pressurized steel piping systems. *Int. J. Press. Vessel. Pip.* **2020**, *180*, 104044. [CrossRef]
4. Garg, R.K.; Kumar, A. Experimental and numerical investigations of water hammer analysis in pipeline with two different materials and their combined configuration. *Int. J. Press. Vessel. Pip.* **2020**, *188*, 104219. [CrossRef]
5. Ostfeld, A. (Ed.) *Water Supply System Analysis—Selected Topics [Internet]*; InTech: London, UK, 2012. [CrossRef]
6. Zhu, M.L.; Shen, B.; Zhang, Y.H.; Wang, T. Research on protection of water hammer in long-distance pressurized water transfer pipeline project. *J. Xi'an Univ. Archit. Technol. (Nat. Sci. Ed.)* **2007**, *39*, 40–43.
7. Zhiyong, L. Comparative research on protective measures of pump-stopping water hammer in water intake pumping station along river. *Trans. Chin. Soc. Agric. Mach.* **2005**, *7*, 61–64.
8. Feng, T.; Jia, Y.; Xie, R. Characteristics and protective measures of water hammer in cascade pumping station for long distance pressure water delivery. *China Water Wastewater* **2008**, *14*, 16.
9. Calamak, M.; Bozkus, Z. Comparison of performance of two run-of-river plants during transient conditions. *J. Perform. Constr. Facil.* **2013**, *27*, 624–632. [CrossRef]
10. Wu, Y.; Xu, Y.; Wang, C. Research on air valve of water supply pipelines. *Procedia Eng.* **2015**, *119*, 884–891. [CrossRef]
11. Arefi, M.H.; Ghaeini-Hessaroeyeh, M.; Memarzadeh, R. Numerical modeling of water hammer in long water transmission pipeline. *Appl. Water Sci.* **2021**, *11*, 140. [CrossRef]
12. Li, G.; Wu, X.; Li, L.; Qiu, W.; Cui, W. Research on water hammer protection for a long-distance water supply system of a deep well pump group. *IOP Conf. Ser. Earth Environ. Sci.* **2021**, *826*, 012047. [CrossRef]
13. Graze, H.R.; Horlacher, H.B. Air Chamber Design charts. In Proceedings of the 8th Australian Fluid Mechanics Conference, University of Newcastle, Newcastle, NSW, Australia, 28 November–2 December October 1982; Volume 28, pp. 14–18.
14. Ruus, E.; Karney, B.W. *Applied Hydraulic Transients*; Ruus Consulting Ltd., Ken Fench: British Columbia, BC, Canada, 1997.
15. Stephenson, D. Simple guide for design of air vessels for water hammer protection of pumping lines. *J. Hydraul. Eng.* **2002**, *128*, 792–797. [CrossRef]
16. Izquierdo, J.; Lopez, P.A.; Lopez, G.; Martinez, F.J.; Perez, R. Encapsulation of air vessel design in a neural network. *Appl. Math. Model.* **2006**, *30*, 395–405. [CrossRef]
17. De Martino, G.; Fontana, N. Simplified approach for the optimal sizing of throttled air chambers. *J. Hydraul. Eng.* **2012**, *138*, 1101–1109. [CrossRef]
18. Sun, Q.; Wu, Y.B.; Xu, Y.; Jang, T.U. Optimal sizing of an air vessel in a long-distance water-supply pumping system using the SQP method. *J. Pipeline Syst. Eng. Pract.* **2016**, *7*, 05016001. [CrossRef]

19. Shi, L.; Zhang, J.; Yu, X.; Chen, S. Water hammer protective performance of a spherical air vessel caused by a pump trip. *Water Supply* **2019**, *19*, 1862–1869. [CrossRef]
20. Wang, X.; Zhang, J.; Yu, X.; Shi, L.; Zhao, W.; Xu, H. Formula for selecting optimal location of air vessel in long-distance pumping systems. *Int. J. Press. Vessel. Pip.* **2019**, *172*, 127–133. [CrossRef]
21. Miao, D.; Zhang, J.; Chen, S.; Li, D.Z. An approximate analytical method to size an air vessel in a water supply system. *Water Sci. Technol. Water Supply* **2017**, *17*, 1016–1021. [CrossRef]
22. Sattar, A.M.; Soliman, M.; El-Ansary, A. Preliminary sizing of surge vessels on pumping mains. *Urban Water J.* **2019**, *16*, 738–748. [CrossRef]
23. Shi, L.; Zhang, J.; Yu, X.D.; Wang, X.T.; Chen, X.Y.; Zhang, Z.X. Optimal volume selection of air vessels in long-distance water supply systems. *AQUA—Water Infrastruct. Ecosyst. Soc.* **2021**, *70*, 1053–1065. [CrossRef]
24. Wan, W.; Zhang, B. Investigation of water hammer protection in water supply pipeline systems using an intelligent self-controlled surge tank. *Energies* **2018**, *11*, 1450. [CrossRef]
25. Charlatte FAYAT Group. A.R.A.A. Dipping Tube Surge Vessel. SPT 140-04-GB. 1978. Available online: https://www.charlattereservoirs.fayat.com (accessed on 17 September 2023).
26. Verhoeven, R.; Van Poucke, L.; Huygens, M. Waterhammer Protection with Air Vessels A Comparative Study. *WIT Trans. Eng. Sci.* **1998**, *18*, 3–14. Available online: https://www.witpress.com/elibrary/wit-transactions-on-engineering-sciences/18/6215 (accessed on 17 September 2023).
27. Ruus, E.; Karney, B.; El-Fitiany, F.A. Charts for water hammer in low head pump discharge lines resulting from water column separation and check valve closure. *Can. J. Civ. Eng.* **1984**, *11*, 717–742. [CrossRef]
28. Boulos, P.F.; Karney, B.W.; Wood, D.J.; Lingireddy, S. Hydraulic transient guidelines for protecting water distribution systems. *J. Am. Water Work. Assoc.* **2005**, *97*, 111–124. [CrossRef]
29. Leruth, P.; Pothof, I. Innovative air vessel design for long distance water transmission pipelines. In *Proceeding of the 11th International Conference on Pressure Surges, Lisbon, Portugal, 24–26 October 2012*; BHR Group Ltd.: Lisbon, Portugal, 2012.
30. Wang, R.; Li, P.; Wang, Z.; Zhang, F. Dipping Tube Hydropneumatic Tank Theory and Application in Water Distribution Systems. In *Proceedings of the 2012 Second International Conference on Electric Technology and Civil Engineering, Washington, DC, USA, 18–20 May 2012*; IEEE Computer Society: Washington, DC, USA, 2012; pp. 958–961. Available online: https://dl.acm.org/doi/proceedings/10.5555/2373291 (accessed on 17 September 2023).
31. Wang, R.H.; Wang, Z.X.; Zhang, F.; Sun, J.L.; Wang, X.X.; Luo, J.; Yang, H.B. Hydraulic transient prevention with dipping tube hydropneumatic tank. *Appl. Mech. Mater.* **2013**, *316*, 762–765. [CrossRef]
32. Moghaddas, S.M.; Samani, H.M.; Haghighi, A. Transient protection optimization of pipelines using air-chamber and air-inlet valves. *KSCE J. Civ. Eng.* **2017**, *21*, 1991–1997. [CrossRef]
33. Hu, J.; Zhai, X.; Hu, X.; Meng, Z.; Zhang, J.; Yang, G. Water Hammer Protection Characteristics and Hydraulic Performance of a Novel Air Chamber with an Adjustable Central Standpipe in a Pressurized Water Supply System. *Sustainability* **2023**, *15*, 29730. [CrossRef]
34. Bentley Systems. Bentley HAMMER Connect. 2023. Available online: https://www.bentley.com (accessed on 17 September 2023).
35. Zhang, Z. *Hydraulic Transients and Computations*, 1st ed.; Springer International Publishing: Cham, Switzerland, 2020.
36. Radi, A.; Poli, R. Genetic programming discovers efficient learning rules for the hidden and output layers of feedforward neural networks. In *Genetic Programming: Second European Workshop, EuroGP'99 Göteborg, Sweden, 26–27 May 1999 Proceedings 2*; Springer Science & Business Media: New York, NY, USA, 1999; pp. 120–134. [CrossRef]
37. Ruus, E. Charts for water-hammer in pipelines with air chambers. *Can. J. Civ. Eng.* **1977**, *4*, 293–313. [CrossRef]
38. Sharif, F.; Siosemarde, M.; Merufinia, E.; Esmatsaatlo, M. Comparative hydraulic simulation of water hammer in transition pipe line systems with different diameter and types. *J. Civ. Eng. Urban.* **2014**, *4*, 282–286, pii: S225204301400043-4.

water

MDPI

Article

MOC-Z Model of Transient Cavitating Flow in Viscoelastic Pipe

Giuseppe Pezzinga

Department of Civil Engineering and Architecture, University of Catania, Via Santa Sofia 64, 95123 Catania, Italy; giuseppe.pezzinga@unict.it

Abstract: In this paper, a unitary method for the solution of transient cavitating flow in viscoelastic pipes is proposed in the framework of the method of characteristics (MOC) and a Z-mirror numerical scheme (MOC-Z model). Assuming a standard form of the continuity equation allows the unified treatment of both viscoelasticity and cavitation. An extension of the MOC-Z is used for Courant numbers less than 1 to overcome a few cases with numerical instabilities. Four viscoelastic models were considered: a Kelvin–Voigt (KV) model without the instantaneous strain, and three generalised Kelvin–Voigt models with one, two, and three KV elements (GKV1, GKV2, and GKV3, respectively). The use of viscoelastic parameters of KV and GKV models calibrated for transient flow tests without cavitation allows good comparisons between experimental and numerical pressure versus time for transient tests with cavitation. Whereas for tests without cavitation, the mean absolute error (MAE) always decreases when the complexity of the model increases (from KV to GKV1, GKV2, and GKV3) for all the considered tests, this does not happen for tests with cavitation, probably because the decreasing capacity of parameter generalization for the increasing complexity of the model. In particular, in the examined cases, the KV model performs better than the GKV1 and the GKV3 models in three cases out of five, and the GKV2 model performs better than the GKV3 model in three cases out of five. Furthermore, the GKV2 model performs better than the KV model only in three cases out of five.

Keywords: transients; cavitation; viscoelasticity; Kelvin–Voigt models; method of characteristics

1. Introduction

In pipe flows, due to high velocities and pipe elevation, the lowering of pressure can give rise to cavities with gas or vapour or both, depending on the fluid flow conditions. When cavities form in transient flows, their collapse can cause pressure rises exceeding the Joukowsky value [1,2]. For pure liquid flow, if the fluid pressure attains its vapor pressure, vaporous cavitation occurs, where the cavities contain only vapor. Transient cavitation in metallic pipes has been deeply studied. A very complete review on the subject was written by Bergant et al. [3]. More rarely has the transient cavitation in polymeric pipes been examined. As a matter of fact, the viscoelastic behaviour of polymeric pipes tends to dampen the transient phenomena, and the effects of transient cavitation in such pipes are less troublesome. But different studies on transient cavitation in viscoelastic pipes exist. The classical experimental tests of Güney [4] were performed for a gravity pipeline of low-density polyethylene (LDPE) for five water temperatures. For each temperature, two tests were performed, without and with cavitation. Other more recent experimental tests of transient cavitating flow in viscoelastic pipes were carried out by Soares et al. [5]. The mentioned tests were used to analyse the behaviour of viscoelastic and cavitation models. The studies on these combined flow effects differ for theoretical and numerical treatment of the viscoelastic behaviour of the pipe and of the cavitation. Another important aspect in which the mentioned studies differ is unsteady friction. Güney [4] solves the continuity and momentum equations by the standard MOC, adopting a GKV model with three Kelvin–Voigt (KV) elements (and seven parameters to calibrate). The cavitation model is a standard discrete vapour cavity model (DVCM) with an additional dissipation term. Unsteady friction is taken into account by the method of the weighting function of Zielke [6] based on a convolution integral as approximated by Trikha [7] for turbulent flow. Hadj-Taïeb and Hadj-Taïeb [8] solve the equations by a two-step Lax-Wendroff [9] scheme and propose a cavitation model in which the mixture density is expressed by a combination of power law in which the choice of a very big exponent generates a quasi-abrupt change in the vapour fraction for pressure equal to the vapour pressure. They neglect unsteady friction and compare the model results with the experimental data of Güney. Soares et al. [5] carry out their own tests and adopt a standard MOC model associated with a DVCM or a discrete gas cavity model (DGCM). The unsteady friction is neglected [5] or taken into account by a modified instantaneous acceleration-based model [10]. Keramat et al. [11] use the experimental data of Soares et al. to test their model based on standard MOC and DVCM. In a more in-depth study, Keramat and Tijsseling [12] also take into account unsteady friction with the Vardy and Brown [13] approximation of the Zielke convolution integral for smooth pipe turbulent flow and fluid–structure interaction. More recently, Urbanowicz and Firkowski [14] use the data of Güney to test their model based on a solution by the MOC of the distributed flow cavitation model proposed by Shu extended to viscoelastic pipes. Unsteady friction is modelled by the convolution integral by Zielke approximated as proposed by Urbanowicz [15]. In subsequent papers, Urbanowicz et al. [16] use the DGCM to model cavitation and Urbanowicz at al. [17] test the discrete Adamkowski cavity model (DACM). Mousavifard [18] propose the use of a two-dimensional (2D) model to study the turbulence in a transient cavitating flow in a viscoelastic pipe, properly allowing for unsteady shear stress computation. The continuity equation combines the form proposed by Pezzinga et al. [19] with regard to the viscoelastic pipe behaviour and the form proposed by Pezzinga and Cannizzaro [20] with regard to the distributed cavitation model. The equations are solved by a finite difference method.

Models of a viscoelastic material usually combine elastic (spring) and viscous (dash-pot) elements. In transient flows, polymeric pipes are widely simulated by the GKV model. The GKV model is made by a purely elastic element (spring), and by one or more KV viscoelastic elements (with spring and dash-pot in parallel). This model can reproduce experimental tests very well [19] if the parameters are calibrated properly. Pezzinga [21] analysed the transient in an installation with a steel pipe and a polymeric additional pipe, finding it difficult to search for general laws of the KV parameters for a GKV model with only one element (three parameters). But if a single element KV model is considered (without instantaneous strain), the viscous parameter of this element, the relaxation time, is significantly related to the oscillation period, which depends on the elastic modulus of the KV element. This correlation was observed also for a whole polymeric pipe [19]. Recently, these results were confirmed with reference to the Güney experiments [22]. Based on these latter results, this paper compares the KV model and GKV models with a different number of elements in tests with cavitation. An implementation is used of the MOC, called MOC-Z, positively tested for transient cavitating pipe flow [23] and for transient flows in viscoelastic pipes [22]. A unified treatment of both viscoelasticity and cavitation in transient pipe flow is proposed. Given that both these aspects influence only the continuity equation, putting this equation in a standard form allows the application of the MOC-Z. Furthermore, a generalization of the numerical method for Courant numbers less than 1 is proposed to overcome possible numerical instabilities. Unsteady friction that can influence the viscous parameter calibration is taken into account by a quasi-two-dimensional (quasi-2D) model. The performances of a KV model and three GKV models with one, two, and three KV elements, identified with GKV1, GKV2, and GKV3, respectively, using the parameters calibrated for tests without cavitation are discussed for tests with cavitation.

2. Mathematical Model

2.1. Continuity and Momentum Equations

The continuity equation can be written in the form:

$$\frac{\partial(\rho A)}{\partial t} + \rho A \frac{\partial u}{\partial x} = 0 \quad (1)$$

if, for sake of simplicity, the convective term $u\partial(\rho A)/\partial x$ is neglected with respect to $\rho A \partial u/\partial x$. In Equation (1), t is the time and ρ is the fluid density, A is the cross-sectional area of the pipe, x is the longitudinal coordinate along the pipe's axis, and u is the velocity component in the longitudinal direction. In the hypothesis of axisymmetric flow, u depends on (x, r, t), r being the distance from the pipe axis, and the fluid density depends only on (x, t). To take into account cavitating flow, assuming vapour cavitation, the fluid density can be assumed as [20]:

$$\rho = \rho_l(1-\alpha) + \rho_v\alpha \quad (2)$$

where ρ_l is the liquid density, ρ_v is the vapour density, and α is the ratio between the volume of vapour and total volume. Given that $\rho_v \ll \rho_l$, Equation (2) can be rewritten in the simplified form [20]:

$$\rho = \rho_l(1-\alpha) \quad (3)$$

By substituting the expression of the density given by Equation (3) in Equation (1), one can write:

$$\frac{\partial p}{\partial t} - \frac{\rho_l a_w^2}{(1-\alpha)}\frac{\partial \alpha}{\partial t} + \rho_l a_w^2 \frac{\partial u}{\partial x} + 2\rho_l a_w^2 \frac{\partial \varepsilon}{\partial t} = 0 \quad (4)$$

where p is the internal pressure, a_w is the wave speed of pure water, and ε is the total circumferential strain of the pipe.

The behaviour of a polymeric pipe can be simulated basically by a KV model with a single KV element. By this assumption, the total strain ε is evaluated by Equation (5) [22]:

$$\frac{\partial \varepsilon}{\partial t} = \frac{1}{\eta}(\sigma - E\varepsilon) \quad (5)$$

with η being the viscosity.

In a GKV model constituted of a simple spring and a series of KV elements [2,5], the total strain is considered the sum of the instantaneous elastic strain ε_i and the retarded viscoelastic strain ε_r, simulated by a series of KV elements [22]:

$$\varepsilon = \varepsilon_i + \varepsilon_r \quad (6)$$

Then, the instantaneous strain is expressed by [22]:

$$\varepsilon_i = \frac{\sigma}{E_0} \quad (7)$$

where E_0 is the instantaneous modulus of elasticity. For a GKV model with n elements, the retarded strain is the sum of the strains of the elements [22]:

$$\varepsilon_r = \sum_{k=1}^{n} \varepsilon_k \quad (8)$$

each of them given by the differential equation [22]:

$$\frac{\partial \varepsilon_k}{\partial t} = \frac{1}{\eta_k}(\sigma - E_k\varepsilon_k) \quad (9)$$

with E_k and η_k being the modulus of elasticity and the viscosity of the k element, respectively. For a circular cross-section pipe, under well-known hypotheses, the Equation (9) relationship can be rewritten in the form [22]:

$$\frac{\partial \varepsilon_k}{\partial t} = \frac{1}{\theta_k}\left(\frac{D\lambda}{2sE_k}p - \varepsilon_k\right) \tag{10}$$

where D is the pipe diameter, s its thickness, λ is a parameter depending on the pipe constraint, and $\theta_k = \eta_k / E_k$ is the retardation time of the k element.

Then, for the GKV model, the quasi-2D continuity equation can be written as:

$$\frac{\partial p}{\partial t} - \frac{\rho_l a_i^2}{(1-\alpha)}\frac{\partial \alpha}{\partial t} + \rho_l a_i^2 \frac{\partial u}{\partial x} + 2\rho_l a_i^2 \sum_{k=1}^{n} \frac{\partial \varepsilon_k}{\partial t} = 0 \tag{11}$$

In Equation (11), a_i is computed by using the instantaneous modulus of elasticity E_i. In the following, three GKV models are considered with one, two, and three KV elements, identified as GKV1, GKV2, and GKV3, respectively.

Once defined, the auxiliary variable φ:

$$\varphi = \frac{p}{\rho_l g} + \frac{a_i^?}{g} ln(1-\alpha) + 2\frac{a_i^2}{g}\sum_{k=1}^{n} \varepsilon_k \tag{12}$$

where g is the gravity acceleration, the continuity equation can be written in the compact form:

$$\frac{\partial \varphi}{\partial t} + \frac{a_i^2}{g}\frac{\partial u}{\partial x} = 0 \tag{13}$$

If a KV model is used, a_i is replaced by a_w, and $n = 1$.

By introducing the further variable:

$$\psi = \varphi - 2\frac{a_i^2}{g}\sum_{k=1}^{n} \varepsilon_k \tag{14}$$

the pressure can be computed as:

$$p = max(\rho_l g \psi, p_v) \tag{15}$$

where p_v is the vapour pressure, and the vapour fraction as:

$$\alpha = max\left[0, 1 - exp\left(\frac{\rho_l g \psi - p_v}{\rho_l a_i^2}\right)\right] \tag{16}$$

The momentum equation of the quasi-2D model is expressed in cylindrical coordinates in the form [22,23]:

$$\frac{\partial u}{\partial t} + g\frac{\partial H}{\partial x} + \frac{1}{\rho r}\frac{\partial (r\tau)}{\partial r} = 0 \tag{17}$$

where τ is the shear stress. Information on the turbulence model can be found in a paper by Pezzinga and Santoro [23].

2.2. Method of Characteristics (MOC)

The application of the MOC to Equations (13) and (17) leads to the equations [23]:

$$\frac{d\varphi}{dt} \pm \frac{a_i}{g}\frac{du}{dt} \pm a_i \frac{\partial (H-\varphi)}{\partial x} \pm \frac{a_i}{\rho_l g r}\frac{\partial (r\tau)}{\partial r} = 0 \tag{18}$$

respectively valid on the characteristic lines:

$$\frac{dx}{dt} = \pm a_i \tag{19}$$

that can be solved numerically.

2.3. Numerical Scheme

The Z-mirror numerical scheme adopted here (Figure 1) was initially proposed for vaporous cavitation transient pipe flow [23] and already successfully applied to transient flow in viscoelastic pipe [22].

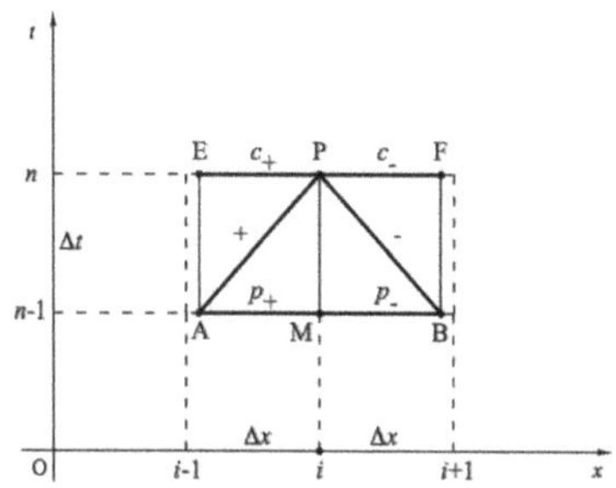

Figure 1. MOC-Z scheme with possible interpolation.

Details on the Z-mirror scheme can be found in the cited references. However, with respect to the previous version of the scheme, here a generalization is proposed to take into account the possibility of assuming a Courant number C less than 1. This extension was motivated because some numerical instabilities were observed in fewer cases when the previously proposed scheme (C equal to 1) was used. Then, the temporal step is computed by the expression:

$$\Delta t = C\frac{\Delta x}{a_i} \tag{20}$$

In the predictor step, Equation (18) is solved as:

$$\frac{\varphi_i^p - \varphi_A}{\Delta t} + \frac{a}{g}\frac{u_{i,j}^p - u_{A,j}}{\Delta t} + a\frac{H_i^{n-1} - \varphi_i^{n-1} - H_A + \varphi_A}{\Delta x} + \frac{2\pi a}{\rho g}\frac{r_{j+1}\tau_{j+1}^{p+} - r_j\tau_j^{p+}}{\Delta A} = 0 \tag{21}$$

$$\frac{\varphi_i^p - \varphi_B}{\Delta t} - \frac{a}{g}\frac{u_{i,j}^p - u_{B,j}}{\Delta t} - a\frac{H_B - \varphi_B - H_i^{n-1} + \varphi_i^{n-1}}{\Delta x} - \frac{2\pi a}{\rho g}\frac{r_{j+1}\tau_{j+1}^{p-} - r_j\tau_j^{p-}}{\Delta A} = 0 \tag{22}$$

The indices i, j, and n refer, to directions x, r, and time t, respectively. When a grid point is used, the previous indices are used, where when a linear interpolated variable is used, this is indicated by the capital letters A, B, E, and F.

In the corrector step, Equation (18) is solved as:

$$\frac{\varphi_i^c - \varphi_A}{\Delta t} + \frac{a}{g}\frac{u_{i,j}^c - u_{A,j}}{\Delta t} + a\frac{H_i^p - \varphi_i^p - H_E + \varphi_E}{\Delta x} + \frac{2\pi a}{\rho g}\frac{r_{j+1}\tau_{j+1}^{c+} - r_j\tau_j^{c+}}{\Delta A} = 0 \tag{23}$$

$$\frac{\varphi_i^c - \varphi_B}{\Delta t} - \frac{a}{g}\frac{u_{i,j}^c - u_{B,j}}{\Delta t} - a\frac{H_F - \varphi_F - H_i^p + \varphi_i^p}{\Delta x} - \frac{2\pi a}{\rho g}\frac{r_{j+1}\tau_{j+1}^{c-} - r_j\tau_j^{c-}}{\Delta A} = 0 \tag{24}$$

The evaluation of the shear stress is carried out using an implicit scheme. The calculations were all carried out with the same numerical grid previously tested for the tests without cavitation [22].

3. Analysis of Results

The model was tested by comparison with the experimental data of Güney [4]. The experimental installation was characterised by an LDPE horizontal gravity pipeline (length L = 43.1 m, diameter D = 0.0416 m, wall thickness s = 0.0042 m) with a reservoir at the upstream end and a valve at the downstream end. Güney performed a couple of experiments, without and with cavitation. respectively, for five different water temperatures. Here, only the tests with cavitation are taken into account, denoted with an even number. In Table 1, for each test, the temperature, the initial velocity, and the upstream reservoir head (considered constant) are reported. The physical properties of water were indirectly deduced from the temperature.

Table 1. Physical parameters of the Güney [4] experiments with cavitation.

Test	T (°C)	V_0 (m/s)	H_s (m)
2	13.8	1.28	2.88
4	25.0	1.37	2.99
6	31.0	1.34	2.98
8	35.0	1.37	2.98
10	38.5	1.33	2.93

The Kelvin–Voigt parameters were calibrated for the Güney tests without cavitation, depending on the temperature [22]. The same parameters, at equal temperature, are used for the tests with cavitation, despite the retardation times depend on the oscillation period and the presence of cavitation influences the temporal development of the phenomenon being shown. The values of the KV parameters used are shown in Table 2.

Table 2. Model parameters [22].

Test		2	4	6	8	10
T (°C)		13.8	25.0	31.0	35.0	38.5
KV	E (GPa)	0.6038	0.4063	0.3321	0.2813	0.2453
	θ (ms)	11.20	15.63	19.16	22.66	25.92
GKV1	E_0 (GPa)	547.8	732.6	583.8	478.8	580.1
	E_1 (GPa)	0.6050	0.4076	0.3351	0.2915	0.2427
	θ_1 (ms)	11.00	15.35	18.64	21.36	25.71
GKV2	E_0 (GPa)	1.221	0.6681	0.5667	0.4099	0.3694
	E_1 (GPa)	1.265	1.436	1.071	1.698	0.9070
	θ_1 (ms)	6.243	5.943	11.38	17.11	17.52
	E_2 (GPa)	2.010	2.038	1.476	1.314	0.9980
	θ_2 (ms)	406.5	130.0	244.9	130.0	292.1
GKV3	E_0 (GPa)	1.652	0.8416	0.7386	0.6641	0.5259
	E_1 (GPa)	1.190	1.193	1.070	0.9610	0.8950
	θ_1 (ms)	4.010	2.207	1.894	2.005	2.057
	E_2 (GPa)	2.325	2.008	1.032	0.8490	0.6470
	θ_2 (ms)	338.3	141.8	279.7	473.0	788.0
	E_3 (GPa)	6.443	5.362	2.439	1.392	0.9841
	θ_3 (ms)	10.76	15.73	19.50	25.84	30.81

Initially, the calculations were carried out with a Courant number equal to 1, as in the standard MOC. However, in some cases, numerical instabilities were observed when the GKV1 model and the GKV3 model were used. To analyse the nature of these instabilities and to confirm their numerical nature, comparisons were made between the results obtained for different Courant numbers. In Figure 2, for Test 6, the numerical results of the GKV3 model for different Courant numbers are compared with the experimental results. No evident differences exist between the results obtained for C = 0.99 (b), C = 0.95 (c), and

C = 0.9 (d) can be noted. The numerical results obtained for C = 1 (a) show instabilities starting after t = 2 s. The dissipative nature of the viscoelastic model allows containment of these instabilities, but the observation that they disappear by lowering the Courant number clearly demonstrates that they are of numerical type.

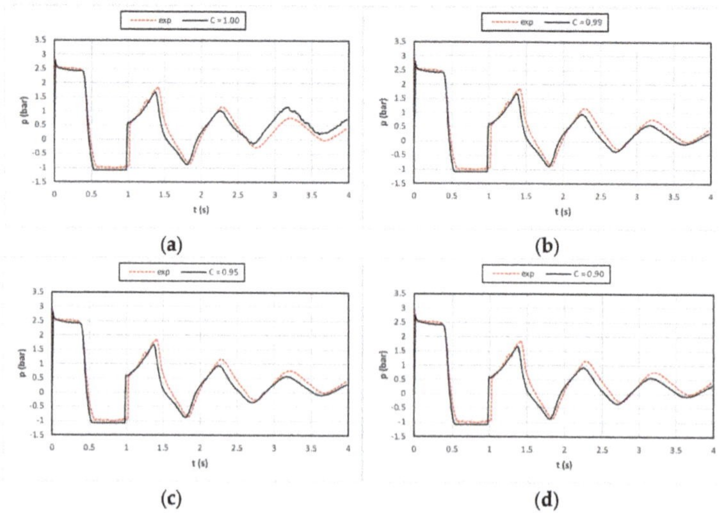

Figure 2. Comparison between numerical and experimental pressure for Test 6 and GKV3 model: (**a**) C = 1, (**b**) C = 0.99, (**c**) C = 0.95, and (**d**) C = 0.9.

In Figure 3, the results of the systematic application of all the experimental tests of the GKV1 model and the GKV3 model with different Courant numbers are summarised in terms of mean absolute error (MAE).

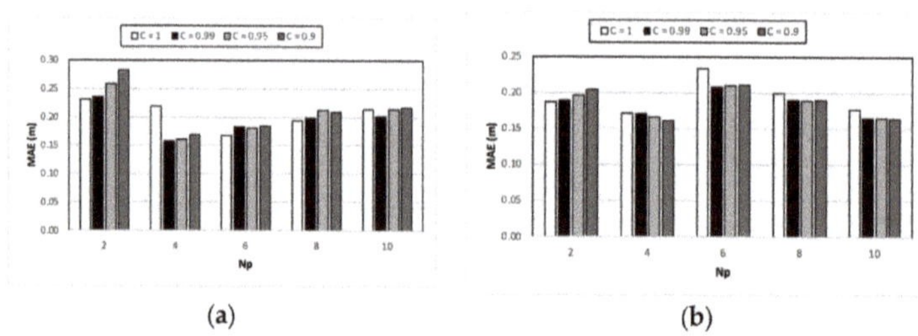

Figure 3. MAE obtained for the GKV1 model (**a**) and for the GKV3 (**b**) model and different Courant numbers.

It can be observed that generally, the Courant number has little influence on the numerical results, but in some cases, where there is a steep change in the MAE, the Courant number is reduced from 1 to 0.99.

Then, for homogeneity, the subsequently presented calculations were obtained with C = 1 for KV and the GKV2 models and with C = 0.99 for the GKV1 and GKV3 models.

In the graphs of Figure 4, the pressure versus time history for the KV, GKV1, GKV2, and GKV3 models is compared with the experimental data for Test 8. The analysis of the results shows that the models with a greater number of parameters (GKV2 and GKV3) reproduce the pressure–time history better in the first part of the transient (about 2 s). But in the second part of the transient (between 2 s and 4 s), the KV model performs better than the GKV models.

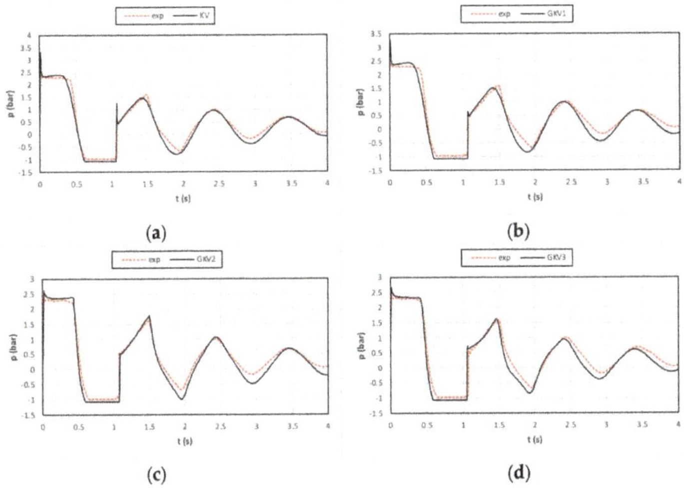

Figure 4. Comparison between numerical and experimental pressure for Test 8: (**a**) KV model, (**b**) GKV1 model, (**c**) GKV2 model, and (**d**) GKV3 model.

As a matter of fact, in the examined cases, there is the problem of generalization of the viscoelastic parameters calibrated for the tests without calibration and extended for the tests with cavitation. Despite the calibrated parameters with equal temperature being used, the cavitation has an influence not only on the pressure but also on the time evolution of the transient. Given that an influence of the oscillation period on the parameters [22] was shown, the calibration should be repeated for the tests with cavitation to obtain the best possible results, but in this way, the effects due to cavitation could be wrongly attributed to viscoelasticity. Then, the method of extending the calibrated parameters for the tests without cavitation to the corresponding ones with cavitation, that is, at the same temperature, seems the best one.

Similar results were obtained for the other tests. The numerical results of the pressure versus time history obtained for tests 2, 4, 6, and 10 by the KV model and by the GKV2 model are compared with the experimental data in Figures 5 and 6, respectively.

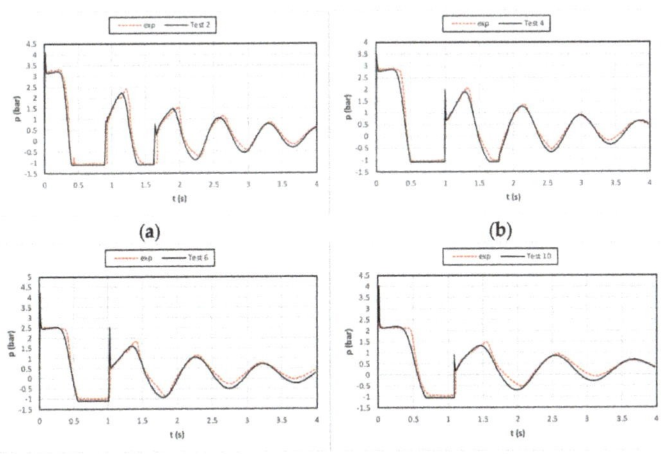

Figure 5. Comparison between numerical and experimental pressure for KV model: (**a**) Test 2, (**b**) Test 4, (**c**) Test 6, and (**d**) Test 10.

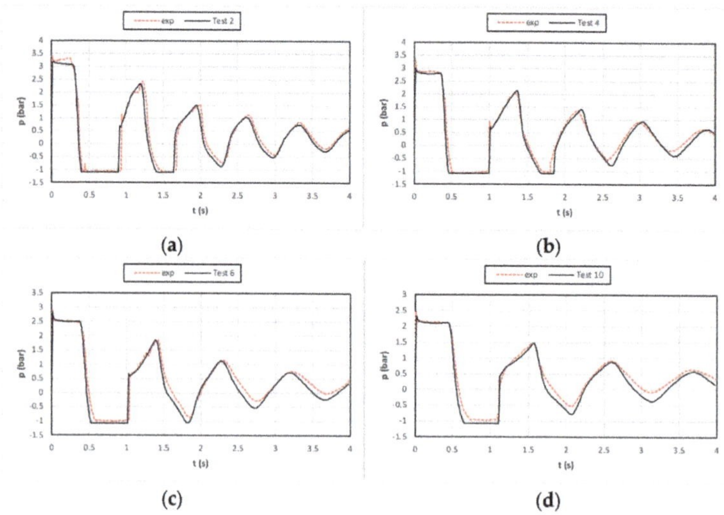

Figure 6. Comparison between numerical and experimental pressure for GKV2 model: (**a**) Test 2, (**b**) Test 4, (**c**) Test 6, and (**d**) Test 10.

The analysis of Figures 5 and 6 allows confirmation of the previous analyses of the influence of the water temperature on the transient. When the temperature increases, the main effect is the increase in deformability, that is, a general decrease in the modulus of elasticity of the KV elements, and the general increase in the retardation time of the KV elements. Then, when the temperature increases, the cavitation episodes are reduced in number and duration. This effect is further illustrated in Figure 7.

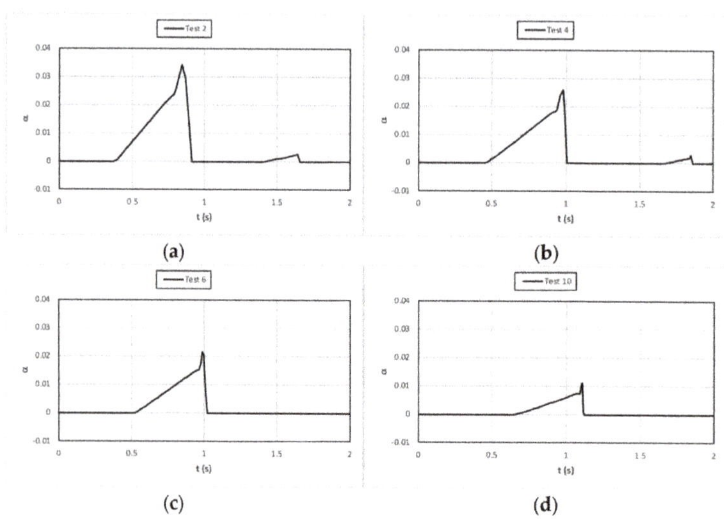

Figure 7. Vapour fraction α at the valve computed by the GKV2 model: (**a**) Test 2, (**b**) Test 4, (**c**) Test 6, and (**d**) Test 10.

In this figure, the vapour fraction α at the valve versus time is reported, as computed by the GKV2 model, for tests 2, 4, 6, and 10. One can see how the presence of the vapour fraction and its maximum values are more and more reduced with increasing temperature.

An analysis of the MAE obtained by the different viscoelastic models for all the examined tests, both without and with cavitation, is presented in Figure 8.

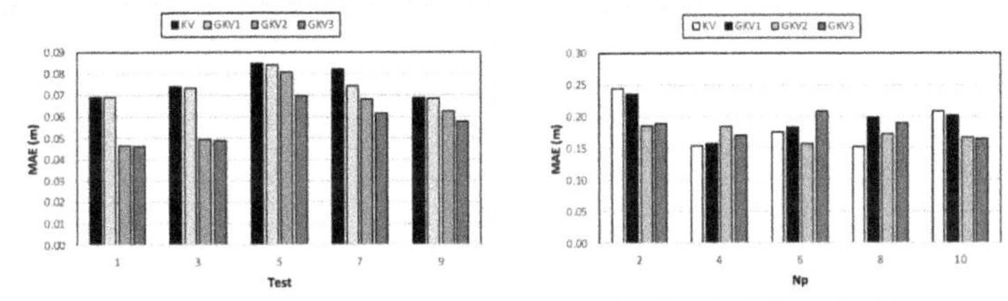

(a) (b)

Figure 8. MAE obtained for the tests without cavitation (**a**) and for the tests with cavitation (**b**).

Analysis of Figure 8 allows some general considerations. As already noted [22], for the tests without cavitation, the increase in the number of KV elements always reduces the MAE, but for almost all the tests, the GKV1 model does not significantly reduce the MAE with respect to the KV model. For the tests with cavitation, the comparison of the values of MAE obtained by the different models shows that in three cases in five, the KV model presents an MAE lower than those obtained by the GKV1 model and the GKV3 model, and in three cases in five, the GKV2 model presents an MAE lower than those obtained by the GKV3 model. This is probably due to the lower capacity of generalization of the calibrated parameters of more complex models with respect to those of simpler ones. The values of MAE obtained by the GKV2 model are lower than those obtained by the KV model only in three cases in five. As such, the GKV2 model performs generally better than the KV model, but not always, as for the tests without cavitation. Therefore, it can be said that the presence of cavitation reduces the differences between different viscoelastic models with respect to the tests without cavitation.

4. Discussion

The analysis of results allows us to make general considerations useful for the hydraulic research and engineering communities. The proposed MOC-Z model for the unified treatment of cavitation and viscoelasticity in transient pipe flow is new, because it allows us to consider, in the frame of the MOC, the most widely used model for transient pipe flow, both the cavitation and the viscoelasticity. In particular, previously studies on cavitation and viscoelasticity usually take into account cavitation by the standard discrete vapour cavity model (DVCM). The only known exception is the model of Mousavifard [18], which unifies the treatment of viscoelasticity and cavitation, but in the frame of a finite difference method.

Another important point is the difference in the present methodology from previous studies. Güney provides the calibrated viscoelastic parameters for his tests for a GKV3 model. All previous researchers that analysed the Güney experiments use these parameters. Here instead, the parameters of the KV, GKV1, GKV2, and GKV3 models calibrated for tests without cavitation were used in order to compare viscoelastic models of different complexity. The experimental data of Güney were used for calibration (tests without cavitation) and validation (tests with cavitation).

Although transient cavitation in viscoelastic pipes is less troublesome than that in elastic pipes, it seems that its treatment in a unitary framework is important, because the proposed approach can be useful in transient pipe flow analysis for each possible similar structure of the continuity equation. Furthermore, the generalization of the MOC-Z for Courant numbers less than 1 allows its use in possible cases in which numerical instabilities arise also.

The extension of the viscoelastic parameters calibrated for tests without cavitation to corresponding tests with cavitation can be questionable because there is a dependence of the KV parameters on the oscillation period, and the presence of cavitation influences the temporal development of the phenomenon.

Probably because of this approach, where in the tests without cavitation, more complex models give rise to decreasing MAE, in the tests with cavitation, this does not generally happen. This different behaviour of the models for the cases without and with cavitation is probably due to the decreasing capacity of parameter generalization for increasing complexity of the models.

On the other hand, a possible calibration of the viscoelastic parameters based on the experimental tests with cavitation could improve the comparisons between numerical and experimental results, but it would not give a general method, because a case-by-case calibration does to seem useful to this end.

The lower capacity of generalization of the calibrated parameters of more complex models with respect to those of simpler ones seems a very hard problem to face. On the other hand, in the scientific literature on transient flow in viscoelastic pipes, there is the tendency to use very complex models with a high number of KV elements just to improve the comparison between numerical and experimental pressure traces. But this is carried out, to the best of the author's knowledge, on the basis of a case-by-case parameter calibration. Recent works of the author on transient flow in viscoelastic pipes endeavour to overcome the lack of generalization of viscoelastic parameters. The correlation of the retardation time with the oscillation period is promising, but it needs further and deeper exploration. In this regard, the presence of cavitation is challenging, because it modifies the time evolution of the transient, making the oscillations only almost periodic until cavitation occurs.

5. Conclusions

The following conclusions can be drawn.

- The analysis of transient cavitation in viscoelastic pipes was carried out by a new mathematical model framing both aspect in the MOC.
- An MOC-Z scheme was adopted and extended to grids with Courant numbers less than 1, then requiring interpolations, to avoid numerical instabilities.
- Viscoelastic models of increasing complexity were adopted, in particular a KV model and three GKV models with one, two, or three KV elements (GKV1, GKV2, and GKV3).
- The analysis of the comparisons between numerical results and experimental data from the literature showed only a limited number of numerical instabilities for the GKV1 and the GKV3 models, for which a Courant number $C = 0.99$ was adopted, capable of eliminating the numerical instabilities.
- The viscoelastic parameters calibrated for the tests without cavitation at corresponding temperature were used, generally showing good agreement between numerical and experimental pressure versus time.
- The KV model gives values of the mean absolute error (MAE) less than those obtained by the GKV1 model and by the GKV3 model in three cases in five and the GKV2 model gives an MAE less than those obtained by the GKV3 model in three cases in five, whereas in the corresponding cases without cavitation, models of increasing complexity give an ever more decreasing MAE, as expected.
- Note that the GKV2 model gives an MAE less than those obtained by the KV model only in three cases in five.

Various aspects could be further improved, such as the flow model and the cavitation model, but the main and most challenging aspect to be deepened is the search for general rules for the evaluation of viscoelastic parameters.

Funding: This research received no external funding.

Data Availability Statement: The numerical results generated during the study are available from the author by request.

Acknowledgments: The author thanks Kamil Urbanowicz for providing the Güney experimental data in digital format.

Conflicts of Interest: The author declares no conflicts of interest.

References

1. Wylie, E.B.; Streeter, V.L. *Fluid Transients in Systems*; Prentice-Hall: Englewood Cliffs, NJ, USA, 1993.
2. Bergant, A.; Simpson, A.R. Pipeline column separation flow regimes. *J. Hydraul. Eng.* **1999**, *125*, 835–848. [CrossRef]
3. Bergant, A.; Simpson, A.R.; Tijsseling, A.S. Water hammer with column separation: A historical review. *J. Fluids Struct.* **2006**, *22*, 135–171. [CrossRef]
4. Güney, M. Contribution à l'étude du Phénomène de Coup de Bélier en Conduite Viscoélastique. Ph.D. Thesis, Université de Lyon I, Lyon, France, 1977. (In French)
5. Soares, A.K.; Covas, D.I.; Carriço, N.J. Transient vaporous cavitation in viscoelastic pipes. *J. Hydraul. Res.* **2012**, *50*, 228–235. [CrossRef]
6. Zielke, W. Frequency-dependent friction in transient pipe flow. *Trans. ASME J. Basic Eng.* **1968**, *901*, 109–115. [CrossRef]
7. Trikha, A.K. An efficient method for simulating frequency-dependent friction in liquid flow. *J. Fluids Eng.* **1975**, *971*, 97–105. [CrossRef]
8. Hadj-Taïeb, L.; Hadj-Taïeb, E. Numerical simulation of transient flows in viscoelastic pipes with vapour cavitation. *Int. J. Modell. Simul.* **2009**, *29*, 206–213. [CrossRef]
9. Lax, P.; Wendroff, B. Systems of Conservation Laws. *Commun. Pure Appl. Math.* **1960**, *13*, 217–237. [CrossRef]
10. Soares, A.K.; Covas, D.I.; Ramos, H.M.; Reis, L.F.R. Unsteady flow with cavitation in viscoelastic pipes. *Int. J. Fluid Mach. Syst.* **2009**, *2*, 269–277. [CrossRef]
11. Keramat, A.; Tijsseling, A.S.; Ahmadi, A. Investigation of transient cavitating flow in viscoelastic pipes. *IOP Conf. Ser. Earth Environ. Sci.* **2010**, *12*, 012081. [CrossRef]
12. Keramat, A.; Tijsseling, A.S. *Waterhammer with Column Separation, Fluid-Structure Interaction and Unsteady Friction in a Viscoelastic Pipe*; CASA-Report; Technische Universiteit Eindhoven: Eindhoven, The Netherlands, 2012; Volume 1243.
13. Vardy, A.E.; Brown, J.M.B. Transient turbulent friction in smooth pipe flows. *J. Sound Vib.* **2003**, *259*, 1011–1036. [CrossRef]
14. Urbanowicz, K.; Firkowski, M. Extended Bubble Cavitation Model to predict water hammer in viscoelastic pipelines. *J. Phys. Conf. Ser.* **2018**, *1101*, 012046. [CrossRef]
15. Urbanowicz, K. Fast and accurate modelling of frictional transient pipe flow. *Z. Angew. Math. Mech.* **2018**, *98*, 802–823. [CrossRef]
16. Urbanowicz, K.; Bergant, A.; Duan, H.F.; Stosiak, M.; Firkowski, M. Using DGCM to predict transient flow in plastic pipe. *IOP Conf. Ser. Earth Environ. Sci.* **2019**, *405*, 012020. [CrossRef]
17. Urbanowicz, K.; Bergant, A.; Duan, H.F. Simulation of unsteady flow with cavitation in plastic pipes using the discrete bubble cavity and Adamkowski models. *IOP Conf. Ser. Mater. Sci. Eng.* **2019**, *710*, 012013. [CrossRef]
18. Mousavifard, M. Turbulence parameters during transient cavitation flow in viscoelastic pipe. *J. Hydraul. Eng.* **2022**, *148*, 04022004. [CrossRef]
19. Pezzinga, G.; Brunone, B.; Meniconi, S. Relevance of pipe period on Kelvin-Voigt viscoelastic parameters: 1D and 2D inverse transient analysis. *J. Hydraul. Eng.* **2016**, *142*, 04016063. [CrossRef]
20. Pezzinga, G.; Cannizzaro, D. Analysis of transient vaporous cavitation in pipes by a distributed 2D model. *J. Hydraul. Eng.* **2014**, *140*, 04014019. [CrossRef]
21. Pezzinga, G. Evaluation of time evolution of mechanical parameters of polymeric pipes by unsteady flow runs. *J. Hydraul. Eng.* **2014**, *140*, 04014057. [CrossRef]
22. Pezzinga, G. On the Characterization of Viscoelastic Parameters of Polymeric Pipes for Transient Flow Analysis. *Modelling* **2023**, *4*, 283–295. [CrossRef]
23. Pezzinga, G.; Santoro, V.C. Shock-Capturing Characteristics Models for Transient Cavitating Pipe Flow. *J. Hydraul. Eng.* **2020**, *146*, 04020075. [CrossRef]

water

MDPI

Article

Development of Pipeline Transient Mixed Flow Model with Smoothed Particle Hydrodynamics Based on Preissmann Slot Method

Yixin Yang [1], Hexiang Yan [1,*], Shixun Li [1], Wenke Song [1,2], Fei Li [1,3], Huanfeng Duan [3], Kunlun Xin [1] and Tao Tao [1]

1 College of Environmental Science and Engineering, Tongji University, Shanghai 200092, China
2 Department of Civil Engineering, The University of Hong Kong, Hong Kong 999077, China
3 Department of Civil and Environmental Engineering, The Hong Kong Polytechnic University, Hong Kong 100872, China
* Correspondence: hxyan@tongji.edu.cn; Tel./Fax: +86-021-6598-5869

Abstract: The accurate modeling and understanding of complex transient mixed pipe flows is crucial for the optimal design and safe and efficient operation in pipeline systems such as urban drainage systems. Currently, the predominant approach for modeling free-surface-pressurized flows relies on grid-based numerical schemes, with comparatively limited capability for exploring its complex phenomena. This study proposed a novel one-dimensional numerical model that integrates the smoothed particle hydrodynamics (SPH) method with the Preissmann slot method (PSM) to explore transient mixed flows in pipeline systems, with better potential capability for exploring more mixed flow phenomena. Empirical parameters of the proposed SPH-PSM model were optimized for improving the numerical accuracy and stability, and the applicable range for these empirical parameters was recommended. The performances of the proposed model were evaluated by different flow regimes, including one free surface case, one fully pressurized case, and two transient mixed-flow cases. The simulation results of different flow regimes demonstrated a high level of agreement with the reference data, indicating the good capability of the SPH-PSM model in simulating complex flow regimes in pipeline systems. Therefore, the proposed SPH-PSM model can be an alternative way for modeling, exploring, and understanding the complex transient mixed flows in pipeline systems.

Keywords: transient mixed flow; Preissmann slot model; smoothed particle hydrodynamics; pipeline systems

1. Introduction

The transient mixed flow is a complicated phenomenon that involves a combination of free surface and pressurized flow regimes. This phenomenon can result in significant damage to the pipeline infrastructure and connected equipment, leading to hydraulic failures in urban drainage systems (UDSs), such as blown-off manhole covers, geysers, and pipe bursts, due to high positive and negative pressure surge fluctuations. Laboratory experiments have been conducted to improve the understanding of the complex flow dynamics involving transition regimes and instability [1–3]. Furthermore, numerical analysis models of transient mixed flow, such as the mouse, hydro-works, and the storm water management model (SWMM) [4–6], were developed to improve the design and operation process in UDSs. However, as pointed out by Bousso et al. [7], the existing numerical models for transient mixed flow have limitations due to the characteristics of multiple phases, transition regimes, and dynamic instabilities (e.g., flow instability, dry bed instability, roll wave instability, free-surface-pressurized flow instability, etc.). Therefore, developing an accurate transient mixed-flow model for UDSs remains a challenging task.

Currently, the one-dimensional numerical model of transient mixed flow in pipes can be categorized into single-phase and two-phase [8]. Single-phase models such as the Preissmann slot method (PSM) and the two-component pressure approach (TPA) model employ one single equation for both free surface and pressurized flows [3,9]. This class of model relaxed the incompressibility to allow additional mass storage. To simulate the negative pressure using PSM, Kerger et al. [10] introduced an original negative PSM model for simulating sub-atmospheric pressurized flow. The advantages of the single-phase models are computational efficiency, simplicity, and unified equations. However, the use of large pressure wave celerity can lead to numerical oscillations at the interface during transitions, and the model's scalability is limited (e.g., in the two-phase flow of water and gas).

Contrary to single-phase models, a two-phase coupled model explicitly considers the gas phase, mainly by the rigid column approach [11], finite-volume strategy [12,13], or fully in the four equations model [14]. Two-phase models incorporate additional equations to describe gas phase and different regimes, allowing for a more comprehensive exploration of the interactions between air and water in mixed flows [1]. And two-phase models could use larger pressure wave celerity in the simulation. However, the implementation of two-phase coupled model requires more complex solving and computational resources [7]; additionally, the code implementation of the model is more complex.

Therefore, considering the practicality of the model, this paper adopted the simple and easily implementable single-equation approach. PSM is a representative transient mixed-flow model and its simulation results are accurate and reliable, with well-established applications in numerous cases. In the latest SWMM5.2 model, PSM was also employed as the transient mixed-flow simulation method [15].

In terms of numerical methods, the traditional numerical schemes used for models described above generally fall into three categories: finite volume method (FVM) [16,17], method of characteristics (MOC) [1,18], and finite difference method (FDM) [2,9]. However, these mesh-based numerical methods still encounter challenges in multiphase transition interface tracking and irregular geometry boundary treatment [7]. On the contrary, the mesh-free methods, such as particle in cell (PIC), smoothed particle hydrodynamics (SPH), etc., have demonstrated significant advantages in dealing with the above-mentioned issues due to their advanced mesh-free features [19]. As a typical mesh-free Lagrangian method, SPH has been rapidly developed due to its excellent potential in modeling fluid dynamics problems characterized by violently changing boundaries or interfaces, complex geometries, and flexibility in tracing moving boundaries. For example, SPH is widely used for modeling dam breaks, sea waves, underwater explosions, and pipe or channel flow modeling [20–22]. For the pipe or channel flow modeling, most SPH studies were developed to solve the open-channel flow [23–25] and the pressurized flow [26,27]. Nevertheless, there is a scarcity of research concerning the application of the SPH method for modeling the transient mixed flow in pipelines [28,29]. Furthermore, mixed flow phenomena are characterized by a dynamic transition interface within a closed conduit, while the SPH method has been proven effective in capturing discrete processes and tracking moving boundaries in transient flows, such as column separation and rapid filling [28]. Therefore, exploring the feasibility and significance of applying the SPH method to solve 1D mixed flow models becomes imperative.

In conclusion, despite the advancements in numerical analysis models for transient mixed flow, the existing numerical models have inadequacies due to their simplicity or the constraints of mesh-based methods. Consequently, the objective of this study is to develop a new numerical solution for modeling transient mixed-flow pipeline systems using the SPH method. Coupled with the PSM model, a SPH-PSM model was proposed and evaluated, focusing on exploring the advanced mesh-free features of SPH in capturing and tracking complex interfaces for pipeline transient mixed flow. The key contributions of this work can be summarized as follows: (1) the development of formulas for the PSM model based on the SPH method; (2) the exploration and optimization of various related parameters; (3) the testing and validation of different flow regimes, including free surface, fully pressurized, and transition mixed flow in pipelines.

2. Numerical Methods

2.1. The Preissmann Slot Model

For a one-dimensional open channel flow, the system of flow governing equations can be written as:

$$\begin{cases} \frac{dA}{dt} = -A\frac{\partial(v)}{\partial x} \\ \frac{dQ}{dt} = -Q\frac{\partial(v)}{\partial x} - g\frac{\partial I}{\partial x} + gA\left(S_0 - S_f\right) \end{cases} \quad (1)$$

where A is the flow area; Q is the flow discharge rate; v is the velocity vector; $S_0 = -z_b/x$ is the bottom slope (z_b is the bottom elevation); and S_f is the friction slope, which is evaluated here according to the Manning formula. The hydrostatic pressure term I is defined as $I = \int_0^h (h-\eta)\sigma(h,\eta)d\eta$, where h is the water depth and $\sigma(h,\eta)$ is the cross-sectional width at an elevation η above the bottom.

According to the Preissmann slot model [30], pressurized flow can equally be calculated through the Saint-Venant equations by adding a conceptual narrow slot on the top of a closed conduit (Figure 1a). When the water level is above the conduit crown, it provides a conceptual free-surface flow, and the slot width T_s should be set according to the relation:

$$T_s = \frac{gA}{a^2} \quad (2)$$

where a is the pressure wave celerity.

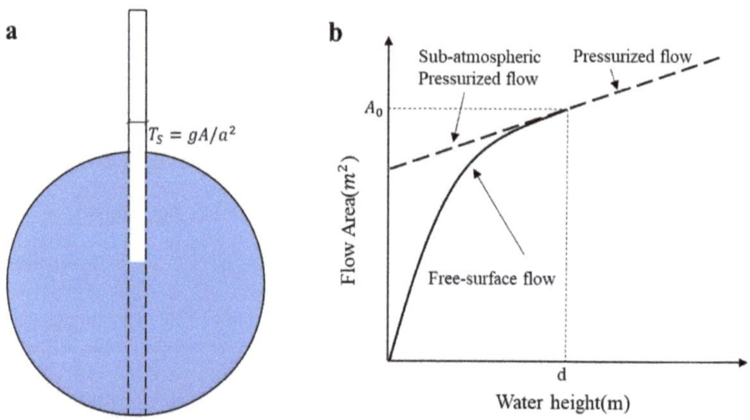

Figure 1. The principles of the negative narrow slot: (**a**) principle schematic; (**b**) relationship bewtween flow area and water height.

To overcome the limitation of the original PSM model in simulating negative pressure, this study adopts the negative narrow slot proposed by Kerger et al. [10]. This approach extends the narrow slot in a straight line below the top of the pipe (Figure 1). The water level below the pipe crown represents two flow area values, one for the free surface flow area and another for the negative pressure flow area. The selection of the appropriate flow area depends on the ventilation conditions.

In the case of a pipe with a circular cross-section, the relevant cross-sectional area can be expressed as follows:

$$A = \begin{cases} A_c + h_s T_s & \text{Pressurized flow} \\ A_c - h_s T_s & \text{Sub-atomspheric pressurized flow} \\ \frac{D^2(\theta - \sin\theta)}{8} & \text{Free-surface flow} \end{cases} \quad (3)$$

where A_C is the cross−sectional area of a circular pipe; h_S is the additional head of the pipe; and θ is the angle between the free surface and the center of the pipe.

2.2. Smoothed Particle Hydrodynamics Method

The SPH method is a complete Lagrangian meshless method, where the entire system is represented by the number of particles with independent masses. The construction of the SPH equation involves two key steps: kernel approximation and particle approximation. Based on this method, the physical property of particle i can be discretized by the following function [31]:

$$\begin{aligned} f(x_i) &= \int_{\Omega} f(x') W(x_i - x', h) dx' \approx \sum_i f(x_j) W(x_i - x_j, h) \frac{m_j}{A_j} \\ \nabla f(x_i) &= \int_{\Omega} f(x') \nabla W(x_i - x', h) dx' \approx \sum_i (f(x_j) - f(x_i)) \nabla W(x_i - x_j, h) \frac{m_j}{A_j} \end{aligned} \tag{4}$$

where $W(x_i - x', h)$ is the kernel function; h is the smoothing length, which determines support domain Ω; m_j is the mass of particle j; A_j is the flow area of particle j; and x_j is the position of particle j.

Although the SPH method has been widely applied in various fields, the traditional SPH method suffers from the boundary defect caused by the truncation of kernel support near the boundaries. To address this drawback, this study employs the corrective SPH (CSPH) method based on Taylor expansion for correction [32,33]. The corrected kernel approximation and gradient are derived as follows:

$$\begin{aligned} \widetilde{W_{ij}} &= \frac{W_{ij}}{\sum_{j=1}^{N} \frac{m_j}{A_j} W_{ij}} \\ \widetilde{\nabla_i W_{ij}} &= \frac{\nabla_i W_{ij}}{\sum_{j=1}^{N} \frac{m_j}{A_j} (x_j - x_i) \cdot \nabla_i W_{ij}} \end{aligned} \tag{5}$$

where $\widetilde{W_{ij}}$ is the corrected kernel function of W_{ij}; and $\widetilde{\nabla_i W_{ij}}$ is the first-order corrected gradient of kernel function $\nabla_i W_{ij}$.

2.3. Discretization of the PSM with SPH

After the aforementioned modifications, the discretized form of PSM, referred to as SPH-PSM, can be approximated as follows:

$$\begin{cases} \frac{DA_i}{Dt} = -\sum_j m_j \left(\frac{Q_j}{A_j} - \frac{Q_i}{A_i}\right) \cdot \widetilde{\nabla W_{ij}} + L_i \\ \frac{DQ_i}{Dt} = -\frac{Q_i}{A_i} \sum_j m_j \left(\frac{Q_j}{A_j} - \frac{Q_i}{A_i}\right) \cdot \widetilde{\nabla W_{ij}} - g A_i \sum_j m_j \left(\frac{I_j}{A_j^2} + \frac{I_i}{A_i^2}\right) \cdot \widetilde{\nabla W_{ij}} \\ \qquad + \sum_j \frac{m_j}{A_j} \Pi_{ij} \cdot \widetilde{\nabla W_{ij}} + g A_i \left(S_{0,i} - S_{f,i}\right) \end{cases} \tag{6}$$

The PSM model uses a detailed integral expression to express I, which is simplified as Equation (7), assuming the pipe is circular.

$$I = \begin{cases} g A_c \left(\frac{D}{2} + h_s\right) + g T_s h_s^2 / 2 & \text{Pressurizedflow} \\ g A_c \left(\frac{D}{2} - h_s\right) + g T_s h_s^2 / 2 & \text{Sub-atomsphericpressurizedflow} \\ \frac{1}{24} g D^3 \left[3 \sin\left(\frac{\theta}{2}\right) \sin^3\left(\frac{\theta}{2}\right) 3 \left(\frac{\theta}{2}\right) \cos\left(\frac{\theta}{2}\right)\right] & \text{Free-surfageflow} \end{cases} \tag{7}$$

The kernel function should have properties such as compact support, normalization, symmetry, and smoothness. Currently, commonly used kernel functions include Gaussian [34], B-spline [35], and quintic-spline kernel functions [36]. The selection of the kernel function in this study will be determined through the model parameter tests in Section 3.1.

To reduce spurious numerical oscillations, the momentum equation is augmented with the normalized artificial viscosity term proposed by Ata et al. [37], which is more suitable for the SPH-SVEs model:

$$\Pi_{ij} = -A_{ij}\frac{\lambda_{ij}(v_i - v_j)(x_i - x_j)}{\sqrt{(x_i - x_j)^2 + \varepsilon^2}} \tag{8}$$

where $A_{ij} = (A_i + A_j)/2$; $\lambda_{ij} = (a_i + a_j)\frac{1}{2}$; and ε is a small constant (chosen as 10^{-6}), to avoid division by zero problem during the computations.

To improve the particle pressure distribution in the SPH method, a density diffusion term is introduced into the continuity equation [38]:

$$L_i = \sum_{j=1}^{N}\frac{m_j}{A_j}\gamma h_{ij}\lambda_{ij}(A_i - A_j)\frac{(x_i - x_j)}{(x_i - x_j)^2 + \varepsilon^2}\widetilde{\nabla_i W_{ij}} \tag{9}$$

where γ is the density diffusion coefficient; and $h_{ij} = 0.5(h_i + h_i)$ is the average smoothing length of the particles.

2.4. Time Integration Scheme

In the SPH method, several time integration methods are commonly used, including the Euler method, leapfrog method [39], and second-order Runge–Kutta (RK2) method [40]. The effects of different time integration methods on the SPH-PSM will be analyzed in Section 3.1. The commonly used Euler method formula is as follows:

$$\boldsymbol{U}^{(n)} = \boldsymbol{U}^{(n-1)} + \boldsymbol{\mathcal{F}}\left(t^{(n-1)}, \boldsymbol{U}^{(n-1)}\right)\cdot\Delta t_{n-1} \tag{10}$$

where $\boldsymbol{U}$ is the conserved variable in the control equation; $\boldsymbol{\mathcal{F}}$ is an expression for the particle approximation of the SPH method; n is the computational step; and Δt_n is the time step at the n.

Additionally, the time step size is set to satisfy the CFL condition:

$$\Delta t \leq \zeta\cdot\min\left(\frac{\Delta x_0}{a_i + |v_i|}\right) \tag{11}$$

where ζ is the Courant number; Δx_0 represents the initial spacing between particles; a_i denotes the pressure wave celerity of particle i; and v_i represents the flow velocity of particle i.

2.5. Boundary Treatments

Accurate and stable modeling relies on the proper treatment of complex boundary conditions. The boundary in SPH methods can be broadly classified into solid wall boundaries and open boundaries.

For solid wall boundaries, a commonly employed method involves introducing mirrored particles [41] to represent the boundaries. For open boundaries, the parameters of buffer particles are determined by the method of characteristics in combination with the energy conservation equation [23,42].

3. Test Results and Discussions

This section shows the impact of the SPH-PSM model parameters on accuracy, stability, and computational efficiency by idealized test cases with exact numerical solutions. Additionally, standard numerical cases and laboratory cases covering three flow regimes (open channel flow, pressurized flow, and transient mixed flow) were utilized to validate the accuracy and versatility of the SPH-PSM model. The simulation codes presented in this section were executed on a computer equipped with an AMD R7 8845H chip and 16 GB of RAM.

3.1. SPH-PSM Model Parameter Optimization

The parameters in the SPH-PSM model can be broadly categorized into two main types: (1) SPH-related parameters: initial particle number n, smoothing length h, kernel function $W(\boldsymbol{x}-\boldsymbol{x}', h)$, density diffusion coefficient γ, time integration scheme, and Courant number ζ; (2) PSM-related parameters: slot width T_s and the pressure wave celerity a.

The discussion of model parameters is based on idealized test cases conducted in a square horizontal pipe with a length of 500 m and dimensions of 1 m × 1 m from Dazzi et al. [43]. The initial discontinuity is located at $x = 250$ m. The test scenario is illustrated in Figure 2. The h_l and v_l represent the piezometric head and flow velocity on the left side, while h_r and v_r represent those variables on the right side. Pipe friction is not considered here. The initial parameters for the three numerical experimental tests are shown in Table 1.

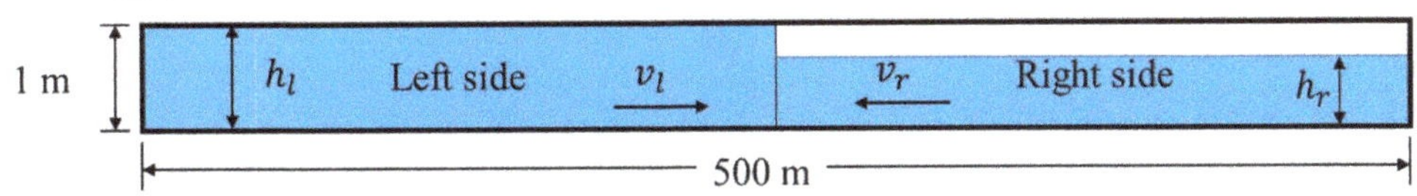

Figure 2. The illustration of the test by Dazzi et al. [43].

Table 1. The initial conditions of the idealized tests from Dazzi et al. [43].

Test Case	h_l (m)	v_l (m/s)	h_r (m)	v_r (m/s)	Study Parameter
T1	3.0	0	0.5	0	SPH-related
T2	1.8	0.9	0.8	−1.0	SPH-related
T3	0.8	2.0	0.8	−2.0	PSM-related

The accuracy of the SPH-PSM model was evaluated by the L_2 norm formula. The reference data are from the numerical simulation results based on the exact Riemann solver proposed by Kerger et al. [17].

$$L_2(\varphi) = \sqrt{\frac{\sum_{i=1}^{N}\left(\varphi_i^{SPH-PSM} - \varphi_i^{REF}\right)^2}{\sum_{i=1}^{N}\left(\varphi_i^{REF}\right)^2}} \tag{12}$$

where the $\varphi_i^{SPH-PSM}$ and φ_i^{REF} mean the value of the simulated results and reference data.

3.1.1. SPH-Related Parameters

The discussion of SPH-related parameters is based on Tests T1 and T2. For reference, a baseline configuration is used with the following settings: a particle number of 500, a smoothing length of $1.3\Delta x_0$, B-spline as the kernel function, the density diffusion coefficient of 0.3, Euler integration as the time integration scheme, a Courant number of 0.5, and the pressure wave celerity is 30 m/s represented by a red dashed line in the figures.

(1) The initial particle number

The initial particle number is a key parameter in the SPH-PSM model, and its selection significantly impacts the simulation results. As shown in Figure 3, it is evident that increasing the initial particle numbers aligns the transition at the discontinuity ($x = 50 \sim 60$ m) with the numerical solution, thereby improving the accuracy of the simulation results. When the number of particles is set to 2000, the model achieves the highest accuracy. However, it is important to note that as the initial particle number increases, the computational time required increases exponentially. Therefore, to achieve the desired accuracy, it is advisable to select a proper number of initial particles based on practical applications.

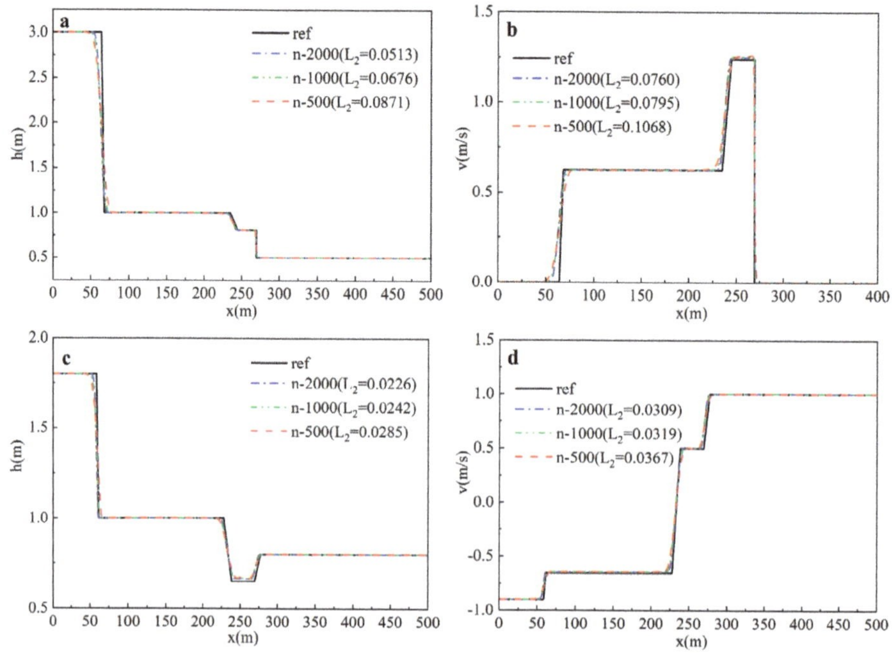

Figure 3. The impact of different particle numbers on test simulation results: (**a**) piezometric head of T1; (**b**) velocity of T1; (**c**) piezometric head of T2; (**d**) velocity of T2.

(2) Smoothing Length

In the SPH method, particle properties are obtained by taking a weighted average with neighboring particles. The smoothing length determines the number of neighboring particles within the support domain of each particle. According to Liu and Moubin [31], this section only compared cases where the smoothing length is greater than $0.5\Delta x_0$.

Figure 4 shows that using a smoothing length between $0.5\Delta x_0$ and $1.0\Delta x_0$ results in higher numerical accuracy, as seen in the pressure simulation results of T1. However, in practical pipelines with pipe slope friction and continuous particle movement, using a small smoothing length can lead to a lack of particles within the support domain, which can impact the computational stability of the program. In the more extreme flow condition T2, the simulation accuracy is lower when the smoothing length is less than 1.

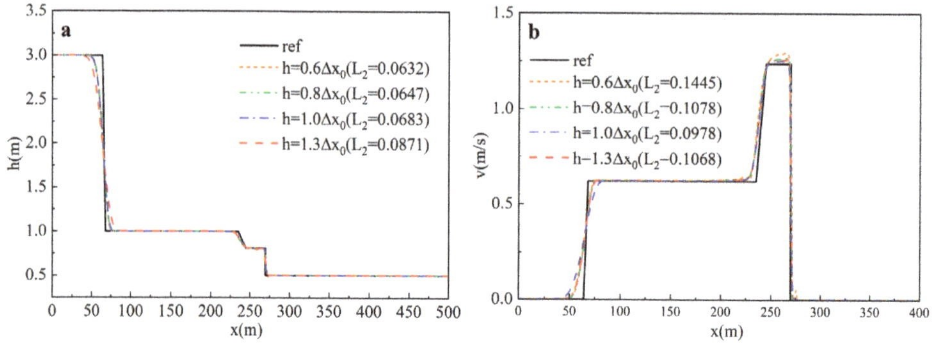

Figure 4. *Cont.*

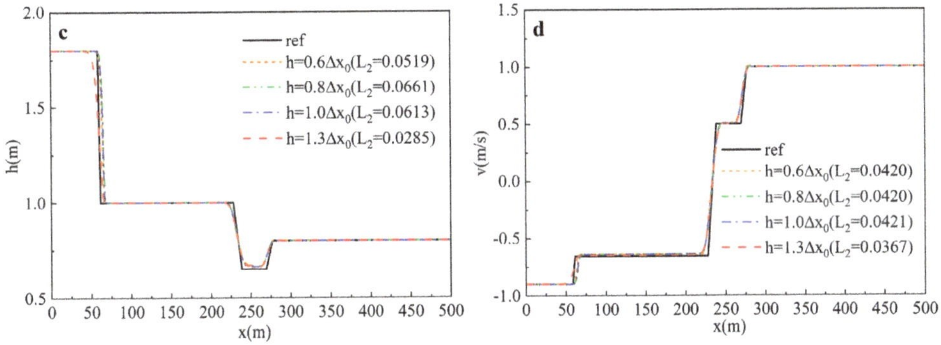

Figure 4. The impact of different smoothing lengths on test simulation results: (**a**) piezometric head of T1; (**b**) velocity of T1; (**c**) piezometric head of T2; (**d**) velocity of T2.

When the smoothing length is greater than $1.0\Delta x_0$, the support domain includes more particles, resulting in smoother results in abrupt transition conditions (e.g., T2). Likewise, in T1, the simulation accuracy decreases at positions where the smoothing length is greater than 1. Therefore, to balance the stability and the accuracy, the smoothing length around $(1.1–1.5)\Delta x_0$ is recommended.

(3) Smoothing Kernel Function

In the SPH-PSM model, the smoothing kernel function determines the weight of the neighboring particles. This study compared the effects of three commonly used smoothing kernel functions–Gaussian, B-spline, and quintic-spline. The accuracy of the B-spline and quintic-spline kernel functions is higher compared to the Gaussian kernel function in the simulation results, as shown in Figure 5. Comparatively, B-spline is widely used in SPH methods due to its high accuracy and concise formula; this study uses B-spline kernel functions for various research cases.

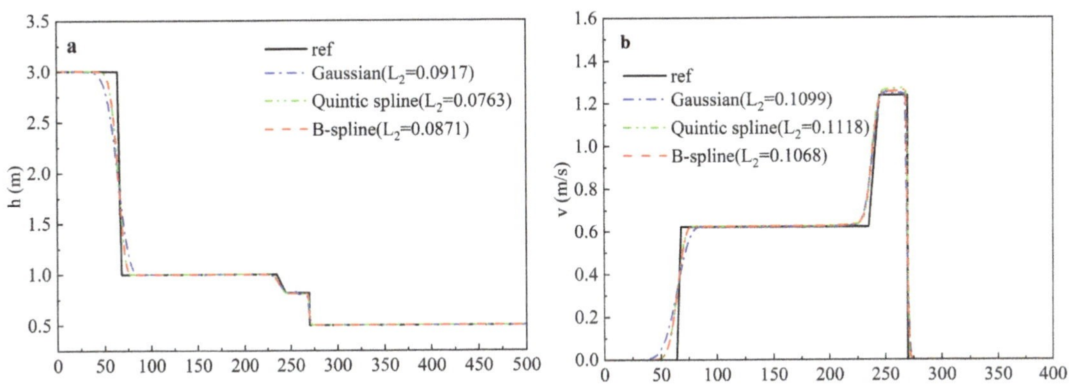

Figure 5. *Cont.*

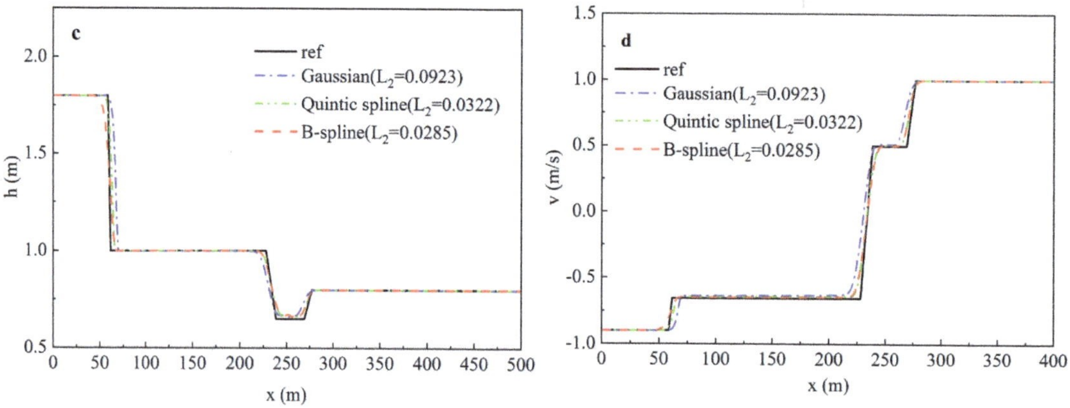

Figure 5. Impact of different smoothing kernel functions on test simulation results: (**a**) piezometric head of T1; (**b**) velocity of T1; (**c**) piezometric head of T2; (**d**) velocity of T2.

(4) Density Diffusion Coefficient

The density diffusion coefficient γ is used to improve the particle pressure distribution in the SPH method. Therefore, it is necessary to choose a reasonable γ during the simulation process. From Figure 6, the influence of the density diffusion coefficient on flow velocity is minor.

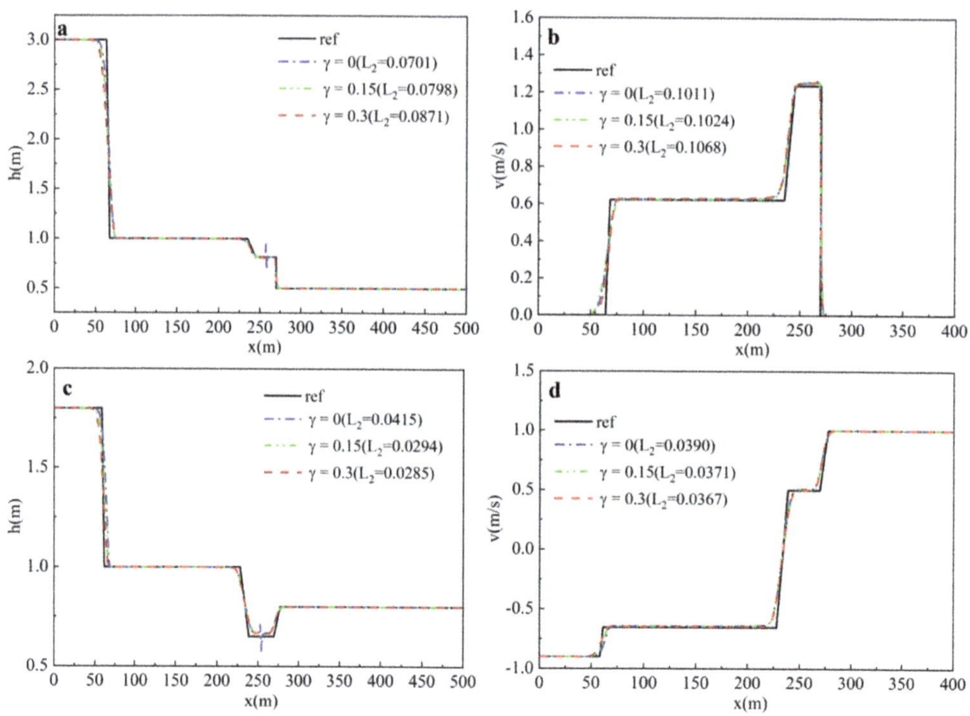

Figure 6. Impact of different density diffusion coefficients γ on test simulation results: (**a**) piezometric head of T1; (**b**) velocity of T1; (**c**) piezometric head of T2; (**d**) velocity of T2.

In Case T1, as the density diffusion coefficient is increased from 0 to 0.3, the pressure numerical oscillations at the initial discontinuity interface (x = 250 m) decrease due to the suppression effect of γ on numerical oscillations from Figure 6. However, this also leads to smoother simulation values in the transition section around 60–70 m, resulting in an overall increase in errors (T1). On the other hand, in Case T2, as the density diffusion coefficient increases, the errors in pressure and velocity gradually decrease. Additionally, the pressure simulation results at the initial discontinuity interface (x = 250 m) also show improvement.

Therefore, the recommended value for the density diffusion coefficient ranges from 0.1 to 0.3 in the SPH-PSM model, while a larger value can be considered in abrupt transition conditions. To ensure stable solving, γ was selected as 0.3 for subsequent research, as it effectively suppresses spurious numerical oscillations at the discontinuity interface while maintaining balance and accuracy.

(5) Time integration scheme

This study discussed the impact of commonly used time integration schemes on the SPH-PSM model, including the Euler method, the leapfrog method, and the second order Runge–Kutta method (RK2).

Based on the findings in Figure 7, it can be observed that the results obtained from the leapfrog method and the RK2 method are quite similar in both cases. However, when it comes to simulating velocity and pressure, the Euler method demonstrates higher accuracy compared to both the leapfrog method and the RK2 method. Furthermore, the Euler method requires less data storage, making it a more efficient choice. Therefore, for the subsequent research in this study, the Euler method was selected as the preferred time integration format.

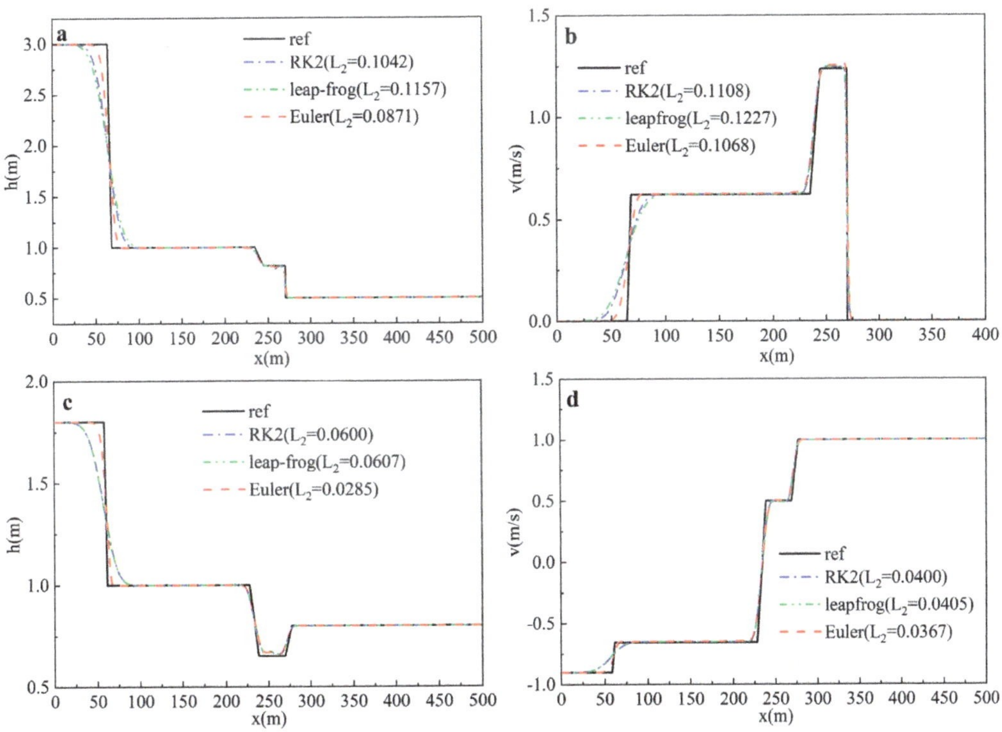

Figure 7. Impact of different time integration formats on test simulation results: (**a**) piezometric head of T1; (**b**) velocity of T1; (**c**) piezometric head of T2; (**d**) velocity of T2.

(6) Courant number

The Courant number (ζ) determines the time step in numerical simulations. For explicit numerical methods, the value of ζ is generally not greater than 1; therefore, this study selected 0.25, 0.5, 0.7, and 0.9 for testing.

From Figure 8, it can be observed that the Courant number has a minor impact on the relatively stable T1 case. As the Courant number increases, the error in piezometric head and velocity even decreases. On the contrary, for the T2 case, when the Courant numbers are greater than 0.5, the simulation result completely deviates from the reference solution of the regime transition interface ($x = 50 \sim 60$ m). This deviation results in gradual increases in errors in pressure and velocity, as shown. The complex hydrodynamic conditions, including opposite flow in T2, require a low Courant number for numerical accuracy and stability. As shown in Table 2, a smaller Courant number leads to longer computational time and higher computational cost. Therefore, to strike a balance between stability, accuracy, and computational cost, it is recommended to use a Courant number (ζ) less than 0.5 for the SPH-PSM model.

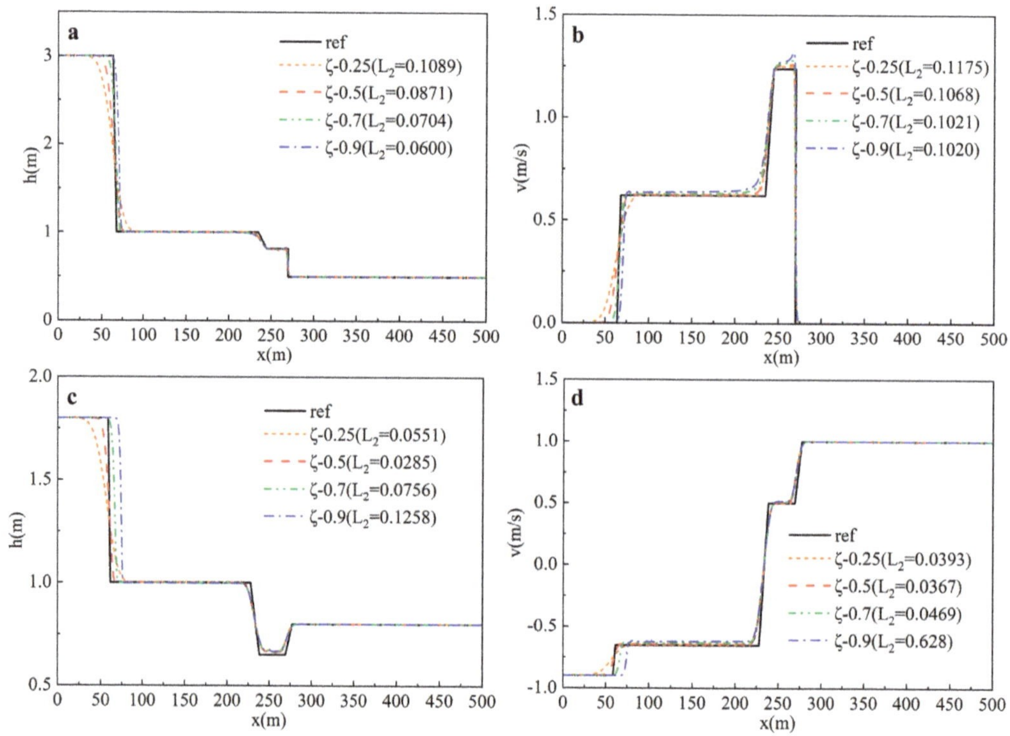

Figure 8. The influence of different Courant numbers on simulation results: (**a**) piezometric head of T1; (**b**) velocity of T1; (**c**) piezometric head of T2; (**d**) velocity of T2.

Table 2. The CPU time for different Courant numbers in various cases.

ζ	0.25	0.5	0.7	0.9
CPU time(s) of T1	4.8606	1.3216	1.0365	0.9318
CPU time(s) of T2	2.9158	0.9700	0.6997	0.6245

3.1.2. PSM-Related Parameters

The parameters related to PSM are the slot width and the pressure wave celerity. However, when the cross-sectional area of the pipeline is determined, T_S and a can be simplified to a single variable parameter according to Equation (2). Generally, to avoid spurious numerical oscillations caused by flow regime transitions, the slot width is usually set to 1–2% of the pipe diameter [2]. Since the slot width is affected by various factors such as the shape of the cross-section, this paper focuses on discussing the pressure wave celerity as an intrinsic influence to ensure the generalizability of the results.

T3 is selected as the benchmark case for the study of PSM-related parameters. In this study, the pressure wave celerity discussed were 99 m/s, 31 m/s, and 15.6 m/s, corresponding to the slot widths of 0.001 m, 0.01 m, and 0.04 m, respectively.

The initial condition of T3 involves two free-surface flows with equal piezometric heads and opposite velocities. When these flows collide, shock waves are generated, resulting in pressurized flows. As the pressure wave celerity increases to 99 m/s, the numerical oscillation becomes apparent and the pressure after collision increases, as shown in Figure 9. Meanwhile, the flow regime transition position moves closer to the upstream and downstream boundaries. Comparing the simulation results in Figure 9, it is clear that as the pressure wave celerity increases, the magnitude of the pressure increases and the advance in the shock position gradually decreases.

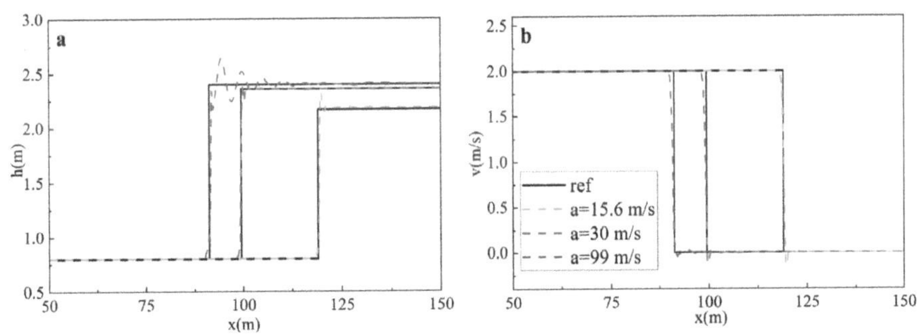

Figure 9. Influence of gap width and wave velocity on simulation with SPH-PSM: (**a**) piezometric head of T3; (**b**) velocity of T3.

When selecting pressure wave celerity, it is necessary to consider the numerical oscillation caused by the narrow slot and the distortion of the simulation results due to the wide slot. The pressure wave celerity of around 30 m/s was found to produce satisfactory simulation results according to Figure 9.

3.2. Numerical Case Validation

In this section, the proposed SPH-PSM model is validated in different flow regimes. The recommended model parameters were used based on the parameter discussion above with a smoothing length of $1.3\Delta x_0$, B-spline as the kernel function, the density diffusion coefficient of 0.3, Euler integration with a Courant number of 0.5, and the pressure wave celerity is 30 m/s. The particle number is determined by the specific case.

3.2.1. Case 1: Dam-Break Case

Dam-break flow cases can be used to test the capability of the SPH-PSM model to simulate single open-channel flow and handle wet–dry interface discontinuities. In this hypothetical example, a horizontal, frictionless fluid with a length of 2000 m is considered. The dam is located at 1000 m, with an initial upstream water depth of 10 m and downstream water depth of 0 m. At $t = 0$, the dam collapses.

For this specific case, initial particle number N was set as 500 (initial spacing Δx_0 is 4 m), while the simulation duration is 60 s. To validate the accuracy of the model, the theoretical solution from Issakhov et al. [44] was utilized as a reference for the dam-break case.

$$\begin{cases} h(x,t) = \frac{1}{9g}\left[2\sqrt{gh_0} - \frac{x}{t}\right]^2 \\ v(x,t) = \frac{2}{3}\left[\sqrt{gh_0} + \frac{x}{t}\right] \end{cases} \tag{13}$$

where h is the water depth; h_0 is the upstream water depth; t is time; x is the coordinate, with positive direction pointing downstream from the dam; v is the average flow velocity at the cross-section of the river; and g is the gravitational acceleration.

Figure 10 illustrates a comparison of the simulation results of water depth and flow velocity in the channel with the analytical solution. As seen in Figure 10, the numerical solution closely aligns with the theoretical solution, indicating the effectiveness of the established dam-break numerical model. These results demonstrate that the SPH-PSM model accurately reproduces the hydraulic conditions of one-dimensional open channels and effectively handles wet–dry interfaces. Overall, this confirms the model's capability for simulating open-channel flow and its potential for widespread use in related applications.

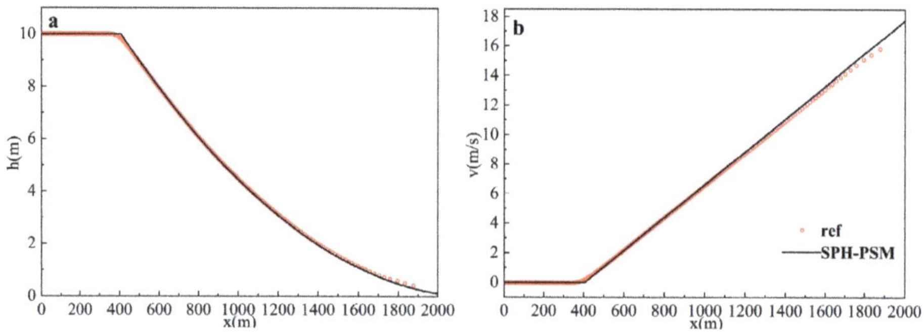

Figure 10. Dry bed dam break case at 60 s: (**a**) water depth; (**b**) flow velocity.

3.2.2. Case 2: Water Hammer Cases

Water hammer cases are conducted to evaluate the SPH-PSM model's ability to simulate pressurized flow and accurately capture negative pressure occurrences in pipelines. In this scenario, a circular pipe with a length of 600 m and a diameter of 0.5 m is considered. Friction in the pipeline is neglected, and the pipe is horizontal. The transient event is triggered by an instantaneous change in the upstream flow rate from 0.477 m^3/s to 0.4 m^3/s, while a downstream water tank maintains a constant water pressure of 45 m.

The initial number of particles was 600 (the initial spacing is 1 m). No regime transition is involved in the water hammer case, so a physically meaningful pressure wave celerity of 1200 m/s was used to improve accuracy, corresponding to a slot width of 1.334×10^{-6} m and the simulation duration was 5 s.

The analytical solution for the water hammer case, as referenced by Kerger et al. [17], predicts a pressure oscillation of approximately 48.05 m following the mentioned change in flow rate. As shown in Figure 11, the simulation results of the SPH-PSM model show good agreement with the analytical solution. The model accurately simulates negative pressure situations and reproduces the propagation process of pressure and flow rate oscillations, including their amplitude, period, and phase.

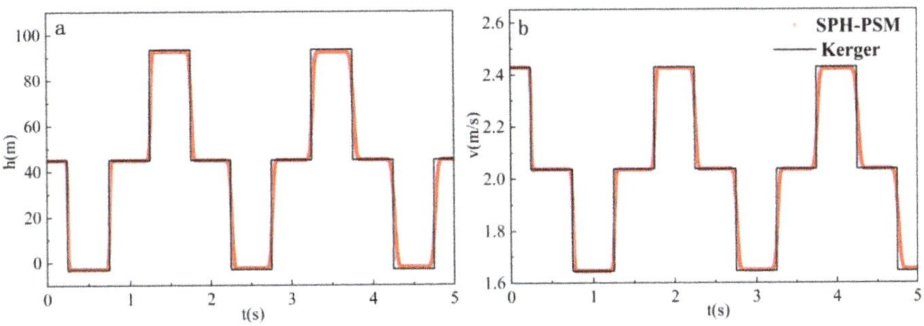

Figure 11. Water hammer case at 300 m from Kerger et al. [17]: (**a**) water height; (**b**) velocity.

3.2.3. Mixed Flow Cases

(1) Case 3: Pipeline Valve Rapid Closure Case

The idealistic test proposed by Dazzi et al. [43] allows for a comparative analysis between SPH-PSM and mesh-based methods. The reference solution of text is based on a fine grid numerical (2000 grids, $\Delta x_0 = 0.05$ m).

The case involves a circular pipe with a length of 100 m, a diameter of 1 m, a slope of 0.4%, and a Manning coefficient of 0.015. The initial flow in the pipe is 0.7 m^3/s, with an initial water level of 0.52 m. The pipe pressurization occurs due to the sudden closure of a downstream valve (at t = 0 s).

The particle numbers were as follows: (a) 100 particles ($\Delta x_0 = 1$ m); (b) 2000 particles ($\Delta x_0 = 0.05$ m), and simulation duration is 60 s. The upstream flow rate is 0.7 m^3/s, and the downstream is a solid wall boundary.

Figure 12 presents the simulation results of the piezometric head at different time steps. Regarding the motion of the free surface and pressurized interface, it can be observed from the figure that, as the downstream valve is closed at t = 0 while normal inflow continues from the upstream, the downstream water level gradually rises to a full pipe flow state. This results in the appearance of a interface between the free surface and the pressurized interface, and with the passage of time, this interface gradually moves upstream. It is evident that the SPH-PSM model successfully simulates the continuous free surface/pressurized interface, accurately describing the movement of the interface and the corresponding pressure.

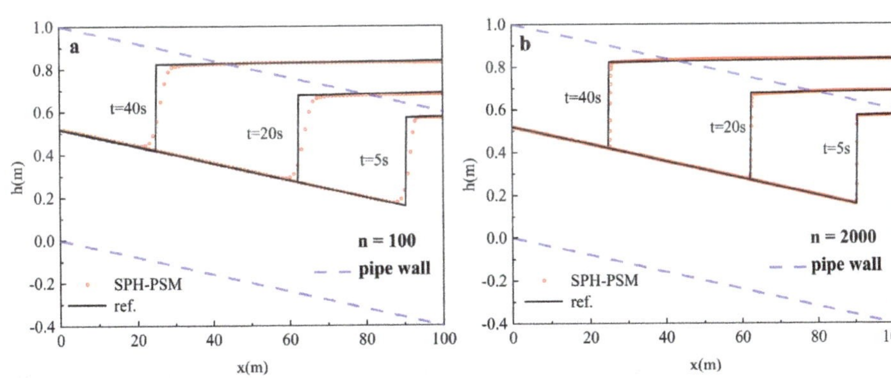

Figure 12. Simulation results of the piezometric head with case from Dazzi et al. [43] in different particle numbers: (**a**) N = 100; (**b**) N = 2000.

When simulating with 100 particles, the pressure at the flow regime transition interface appears to be relatively smooth. The simulation results with 100 particles can also accurately reflect the motion of free surfaces, pressurized interface, and the water level after pressurization in the model. Increasing the number of particles to 2000 leads to a more accurate simulation that better captures the abrupt change in pressure at the flow regime transition interface. From the results, it is evident that a finer particle subdivision yields more accurate simulation results. This finding is consistent with the analysis presented in Section 3.1.1.

(2) Case 4: Single Drainage Pipe Case

The experimental setup designed by Vasconcelos et al. [3] simulates a simplified "stormwater well-pipe-stormwater well" system. In this case, water continuously flows into the upstream, and both upstream and downstream are open boundaries. The method of characteristics is employed to solve boundary conditions.

As shown in Figure 13, the experimental setup consists of an upper and lower stormwater well connected by a pipe. The horizontal pipe is 14.33 m long with a diameter of 0.094 m. The upstream stormwater well measures 0.31 m in height and has a square bottom with dimensions of 0.25 m by 0.25 m. The down stormwater well is designed with a circular bottom, with a height sufficient to prevent water overflow and a bottom diameter of 0.19 m. The Manning coefficient is 0.012. A gate located downstream of the pipe is installed. To eliminate the impact of trapped air within the system on the experiment, a ventilation pipe is arranged near the gate.

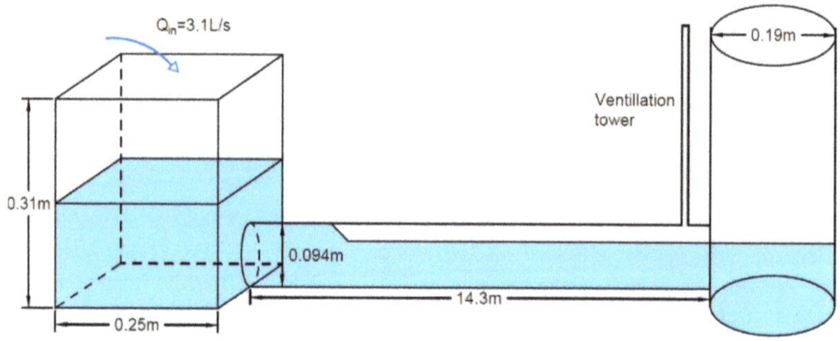

Figure 13. Conceptual diagram of the experimental setup from Vasconcelos et al. [3].

The initial water depth in the experiment is 0.073 m and a constant inflow of 3.1 L/s is set at the upstream rainwater well. Water exceeding the height of the upstream rainwater well will overflow from the top of the pipe. Pressure and velocity data were measured 9.9 m away from the upstream boundary. The initial spacing of particles is 0.1 m. The simulation duration is 40 s.

It is worth noting that the experiment described a partially closed gate at the downstream end of the pipeline. This partial closure may restrict water flow, thereby reducing the observed oscillations in the experiment and slowing down the propagation speed of pressure waves upstream. Therefore, we modified the model by adding a local head loss term h_f to roughly account for the gate's influence, as follows:

$$h_f = (\xi)^2 \frac{v^2}{2g} \tag{14}$$

As per the reference literature [45], when the local head loss coefficient ξ = 25, the modified simulation results achieved better consistency.

As shown in Figure 14, the phases of the oscillations are essentially the same, but the model underpredicts the damping of the subsequent oscillations. These discrepancies may be attributed to an underestimation of energy losses [3] during the simulation process or constant pressure wave celerity [15]. Although there are differences in pressure simulation, considering that existing models also exhibit such differences, we deem this difference acceptable [3,38,46].

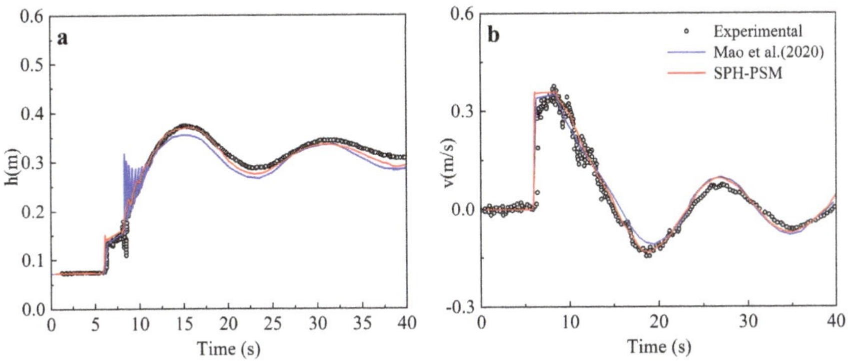

Figure 14. The proposed SPH-PSM model results at 9.9 m comparison with the laboratory-observed data from Vasconcelos et al. [3]. and numerical results from Mao et al. [46]: (**a**) flow piezometric head; (**b**) flow velocity.

Overall, the SPH-PSM model can effectively reproduce satisfying results of the classical numerical and experimental cases when compared to the reference data from the literature. The recommended optimal parameters of the proposed model prove to be effective in addressing different flow conditions.

4. Conclusions

In this study, a one-dimensional pipeline transient mixed-flow model called the SPH-PSM model is established by combining smoothed particle hydrodynamic methods and the Preissmann slot method. Subsequently, the empirical parameters within the SPH-PSM model are systematically tested and discussed. Finally, the proposed model with optimal parameters is validated with classic numerical and experimental cases from the literature. The key findings are summarized as follows:

(1) The newly proposed SPH-PSM model demonstrates its effectiveness in accurately simulating complex transient flow regimes.
(2) The recommended model parameters for SPH-PSM are specified as follows: (a) the B-spline kernel function is more appropriate in the SPH-PSM model; (b) the recommended smoothing length is (1.1–1.5)Δx_0 to balance stability and accuracy; (c) the density diffusion coefficient γ can suppress the numerical oscillations while potentially compromising mass conservation, which is recommended as 0.1–0.3, while a larger γ is recommended for intense mixed flows; (d) among different numerical integration methods, the Euler method proves to be superior in terms of computation time and accuracy; (e) to maintain a balance between stability, accuracy, and efficiency, a Courant number less than 0.5 is recommended; (f) the choice of pressure wave celerity affects not only the numerical oscillation but also the position of the shock and the piezometric head, which is recommended as 30 m/s; meanwhile, in the pure pressurized flow, the true acoustic velocity should be chosen to ensure accuracy.

In conclusion, the proposed SPH-PSM model exhibits its versatility in simulating transient mixed flows within pipelines. It provides an alternative approach to modeling pipe fluid dynamics. Attributable to the mesh-free nature of SPH, the SPH-PSM model can be further extended to solve multiphase flow issues like entrapped air pockets and sediment transport in future research.

Author Contributions: All authors contributed to the study conception and design. Material preparation, data collection, and analysis were performed by Y.Y., S.L., W.S. and H.Y.; the first draft of the manuscript was written by Y.Y. and revised by H.Y., H.D., F.L., W.S., K.X. and T.T.; project fundings were supported by K.X., H.Y., F.L. and T.T. All authors commented on previous versions of the manuscript. All authors have read and agreed to the published version of the manuscript.

Funding: This work was supported by the Shanghai Science and Technology Innovation Action Plan [grant No. 22dz1201800], and the National Natural Science Foundation of China [grant numbers: 51808396, 52000142, 51978493].

Data Availability Statement: All data generated or analyzed during this study are included in this published article.

Acknowledgments: We sincerely thank the editors and reviewers for their valuable comments during the review process of the paper.

Conflicts of Interest: The authors declare no competing interests.

Abbreviations

Latin letters:

a	Pressure wave celerity
A	Flow area
A_c	Cross-sectional area
D	Pipe inner diameter
g	Gravitational acceleration
h	Smoothing length
h_S	Additional head of the pipe
h_f	Local head loss term
I	Hydrostatic pressure term
L_i	Density diffusion term of particle i
$L_2(\varphi)$	Accuracy evaluate norm formula
m_i	Mass of particle i
n	The initial particle number
Q	Flow discharge rate
S_0	The bottom slope
S_f	Friction slope
t	Time
T_s	Slot width
v	Velocity vector
W_{ij}	Kernel function
$\nabla_i W_{ij}$	The first-order gradient of kernel function
$\widetilde{W_{ij}}$	Corrected kernel function
$\widetilde{\nabla_i W_{ij}}$	The first-order corrected gradient of kernel function
x_i	Position of particle i
Δx_0	Initial spacing between particles
z_b	Bottom elevation

Greek letters:

θ	Angle between the free surface and the center of the pipe
γ	The density diffusion coefficient
Π_{ij}	Artificial viscosity
ζ	Courant number
ξ	Local head loss term coefficient

Acronyms:

CFL	Courant–Friedrichs–Lewy condition
CSPH	The Corrective SPH
FDM	Finite Difference Method
FVM	Finite Volume Method
MOC	Method of Characteristics
PIC	Particle in cell
PSM	The Preissmann Slot Method
RK2	The Second Order Runge–Kutta method
SPH	The Smoothed Particle Hydrodynamics
UDSs	Urban drainage systems

References

1. Aureli, F.; Dazzi, S.; Maranzoni, A.; Mignosa, P. Validation of single- and two-equation models for transient mixed flows: A laboratory test case. *J. Hydraul. Res.* **2015**, *53*, 440–451. [CrossRef]
2. Trajkovic, B.; Ivetic, M.; Calomino, F.; D'Ippolito, A. Investigation of transition from free surface to pressurized flow in a circular pipe. *Water Sci. Technol.* **1999**, *39*, 105–112. [CrossRef]
3. Vasconcelos, J.G.; Wright, S.J.; Roe, P.L. Improved simulation of flow regime transition in sewers: Two-component pressure approach. *J. Hydraul. Eng.* **2006**, *132*, 553–562. [CrossRef]
4. Chitwatkulsiri, D.; Miyamoto, H. Real-Time Urban Flood Forecasting Systems for Southeast Asia—A Review of Present Modelling and Its Future Prospects. *Water* **2023**, *15*, 178. [CrossRef]
5. Elliott, A.H.; Trowsdale, S.A. A review of models for low impact urban stormwater drainage. *Environ. Modell. Softw.* **2007**, *22*, 394–405. [CrossRef]
6. Farina, A.; Di Nardo, A.; Gargano, R.; van der Werf, J.A.; Greco, R. A simplified approach for the hydrological simulation of urban drainage systems with SWMM. *J. Hydrol.* **2023**, *623*, 129757. [CrossRef]
7. Bousso, S.; Daynou, M.; Fuamba, M. Numerical Modeling of Mixed Flows in Storm Water Systems: Critical Review of Literature. *J. Hydraul. Eng.* **2013**, *139*, 385–396. [CrossRef]
8. Li, F.; Yan, J.; Yan, H.; Tao, T.; Duan, H.-F. 2D Modelling and energy analysis of entrapped air-pocket propagation and spring-like geysering in the drainage pipeline system. *Eng. Appl. Comput. Fluid Mech.* **2023**, *17*, 2227662. [CrossRef]
9. Cunge, J.A.; Wegner, M. Numerical integration of Barré de Saint-Venant's flow equations by means of an implicite scheme of finite differences. *La Houille Blanche* **1964**, *50*, 33–39. [CrossRef]
10. Kerger, F.; Archambeau, P.; Erpicum, S.; Dewals, B.J.; Pirotton, M. A fast universal solver for 1D continuous and discotinuous steady flows in rivers and pipes. *Int. J. Numer. Methods Fluids* **2011**, *66*, 38–48. [CrossRef]
11. Li, J.; McCorquodale, A. Modeling mixed flow in storm sewers. *J. Hydraul. Eng.* **1999**, *125*, 1170–1180. [CrossRef]
12. Huang, B.; Zhu, D.Z. Rigid-column model for rapid filling in a partially filled horizontal pipe. *J. Hydraul. Eng.* **2021**, *147*, 06020018. [CrossRef]
13. An, H.; Lee, S.; Noh, S.J.; Kim, Y.; Noh, J. Hybrid Numerical Scheme of Preissmann Slot Model for Transient Mixed Flows. *Water* **2018**, *10*, 899. [CrossRef]
14. Henau, V.D.; Raithby, G. A transient two-fluid model for the simulation of slug flow in pipelines—II. Validation. *Int. J. Multiph. Flow* **1995**, *21*, 351–364. [CrossRef]
15. Sharior, S.; Hodges, B.; Vasconcelos, J.G. Generalized, Dynamic, and Transient-Storage Form of the Preissmann Slot. *J. Hydraul. Eng.* **2023**, *149*, 04023046. [CrossRef]
16. León, A.S.; Ghidaoui, M.S.; Schmidt, A.R.; García, M.H. Godunov-type solutions for transient flows in sewers. *J. Hydraul. Eng.* **2006**, *132*, 800–813. [CrossRef]
17. Kerger, F.; Archambeau, P.; Erpicum, S.; Dewals, B.J.; Pirotton, M. An exact Riemann solver and a Godunov scheme for simulating highly transient mixed flows. *J. Comput. Appl. Math.* **2011**, *235*, 2030–2040. [CrossRef]
18. Chaudhry, M.H. *Open-Channel Flow*; Springer: Berlin/Heidelberg, Germany, 2008; Volume 523.
19. Ye, T.; Pan, D.; Huang, C.; Liu, M. Smoothed particle hydrodynamics (SPH) for complex fluid flows: Recent developments in methodology and applications. *Phys. Fluids* **2019**, *31*, 011301. [CrossRef]
20. Monaghan, J.J. Simulating free surface flows with SPH. *J. Comput. Phys.* **1994**, *110*, 399–406. [CrossRef]
21. Swegle, J.; Attaway, S. On the feasibility of using smoothed particle hydrodynamics for underwater explosion calculations. *Comput. Mech.* **1995**, *17*, 151–168. [CrossRef]
22. Salis, N.; Luo, M.; Reali, A.; Manenti, S. Wave generation and wave-structure impact modelling with WCSPH. *Ocean Eng.* **2022**, *266*, 113228. [CrossRef]
23. Chang, T.J.; Chang, K.H. SPH Modeling of One-Dimensional Nonrectangular and Nonprismatic Channel Flows with Open Boundaries. *J. Hydraul. Eng.* **2013**, *139*, 1142–1149. [CrossRef]
24. Chang, K.H.; Chang, T.J. A well-balanced and positivity-preserving SPH method for shallow water flows in open channels. *J. Hydraul. Res.* **2021**, *59*, 903–916. [CrossRef]
25. Pan, T.W.; Zhou, L.; Ou, C.Q.; Wang, P.; Liu, D.Y. Smoothed Particle Hydrodynamics with Unsteady Friction Model for Water Hammer Pipe Flow. *J. Hydraul. Eng.* **2022**, *148*, 04021057. [CrossRef]
26. Wieth, L.; Kelemen, K.; Braun, S.; Koch, R.; Bauer, H.J.; Schuchmann, H.P. Smoothed Particle Hydrodynamics (SPH) simulation of a high-pressure homogenization process. *Microfluid. Nanofluid* **2016**, *20*, 42. [CrossRef]
27. Zhou, M.J.; Shi, Z.M.; Peng, C.; Peng, M.; Cui, K.F.E.; Li, B.; Zhang, L.M.; Zhou, G.G.D. Two-phase modelling of erosion and deposition process during overtopping failure of landslide dams using GPU-accelerated ED-SPH. *Comput. Geotech.* **2024**, *166*, 105944. [CrossRef]
28. Nomeritae, N.; Bui, H.H.; Daly, E. Modeling Transitions between Free Surface and Pressurized Flow with Smoothed Particle Hydrodynamics. *J. Hydraul. Eng.* **2018**, *144*, 04018012. [CrossRef]
29. Song, W.K.; Yan, H.X.; Tao, T.; Guan, M.F.; Li, F.; Xin, K.L. Modeling Transient Mixed Flows in Drainage Networks With Smoothed Particle Hydrodynamics. *Water Resour. Manag.* **2024**, *38*, 861–879. [CrossRef]
30. Preissmann, A. *Propagation des Intumescences Dans Lescanaux et Rivieres*; 1st Cong. French Assoc. for Computation: Grenoble, France, 1961.
31. Liu, G.-R.; Liu, M.B. *Smoothed Particle Hydrodynamics: A Meshfree Particle Method*; World Scientific: Singapore, 2003.

32. Bonet, J.; Lok, T.S.L. Variational and momentum preservation aspects of Smooth Particle Hydrodynamic formulations. *Comput. Methods Appl. Mech. Eng.* **1999**, *180*, 97–115. [CrossRef]
33. Chen, J.K.; Beraun, J.E.; Carney, T.C. A corrective smoothed particle method for boundary value problems in heat conduction. *Int. J. Numer. Methods Eng.* **1999**, *46*, 231–252. [CrossRef]
34. Monaghan, J.J.; Gingold, R.A. Shock Simulation by the Particle Method Sph. *J. Comput. Phys.* **1983**, *52*, 374–389. [CrossRef]
35. Lahiri, S.K.; Bhattacharya, K.; Shaw, A.; Ramachandra, L.S. A stable SPH with adaptive B-spline kernel. *J. Comput. Phys.* **2020**, *422*, 109761. [CrossRef]
36. Yang, X.F.; Peng, S.L.; Liu, M.S. A new kernel function for SPH with applications to free surface flows. *Appl. Math. Model.* **2014**, *38*, 3822–3833. [CrossRef]
37. Ata, R.; Soulaïmani, A. A stabilized SPH method for inviscid shallow water flows. *Int. J. Numer. Methods Fluids* **2005**, *47*, 139–159 [CrossRef]
38. Chang, K.H.; Sheu, T.W.H.; Chang, T.J. A 1D-2D coupled SPH-SWE model applied to open channel flow simulations in complicated geometries. *Adv. Water Resour.* **2018**, *115*, 185–197. [CrossRef]
39. Vacondio, R.; Rogers, B.; Stansby, P. Smoothed Particle Hydrodynamics: Approximate zero-consistent 2-D boundary conditions and still shallow-water tests. *Int. J. Numer. Methods Fluids* **2012**, *69*, 226–253. [CrossRef]
40. Mabssout, M.; Herreros, M.I. Runge-Kutta vs Taylor-SPH: Two time integration schemes for SPH with application to Soil Dynamics. *Appl. Math. Model.* **2013**, *37*, 3541–3563. [CrossRef]
41. Federico, I.; Marrone, S.; Colagrossi, A.; Aristodemo, F.; Antuono, M. Simulating 2D open-channel flows through an SPH model. *Eur. J. Mech.-B/Fluids* **2012**, *34*, 35–46. [CrossRef]
42. Song, W.K.; Yan, H.X.; Li, F.; Tao, T.; Duan, H.F.; Xin, K.L.; Li, S.P. Development of Smoothed Particle Hydrodynamics based water hammer model for water distribution systems. *Eng. Appl. Comput. Fluid Mech.* **2023**, *17*, 2171139. [CrossRef]
43. Dazzi, S.; Maranzoni, A.; Mignosa, P. Local time stepping applied to mixed flow modelling. *J. Hydraul. Res.* **2016**, *54*, 145–157. [CrossRef]
44. Issakhov, A.; Zhandaulet, Y.; Abylkassymova, A. Numerical Study of the Water Surface Movement during a Dam Break on a Slope with Cascade Dike from Sediment. *Water Resour. Manag.* **2022**, *36*, 3435–3461. [CrossRef]
45. Sanders, B.F.; Bradford, S.F. Network implementation of the two-component pressure approach for transient flow in storm sewers. *J. Hydraul. Eng.* **2011**, *137*, 158–172. [CrossRef]
46. Mao, Z.H.; Guan, G.H.; Yang, Z.H. Suppress Numerical Oscillations in Transient Mixed Flow Simulations with a Modified HLL Solver. *Water* **2020**, *12*, 1245. [CrossRef]

water

MDPI

Article

Smoothed Particle Hydrodynamics Simulations of Porous Medium Flow Using Ergun's Fixed-Bed Equation

Carlos E. Alvarado-Rodríguez [1,2,*,†], Lamberto Díaz-Damacillo [2,3,†], Eric Plaza [4,†] and Leonardo Di G. Sigalotti [3,*,†]

1 Departamento de Ingeniería Quimica, DCNE, Universidad de Guanajuato, Noria Alta S/N, Guanajuato 03605, Mexico
2 Consejo Nacional de Ciencia y Tecnología, Avenida Insurgentes Sur 1582, Crédito Constructor, Ciudad de México 03940, Mexico; diaz.lamberto@gmail.com
3 Departamento de Ciencias Básicas, Universidad Autónoma Metropolitana-Azcapotzalco (UAM-A), Av. San Pablo 420, Colonia Nueva el Rosario, Alcaldía Azcapotzalco, Ciudad de México 02128, Mexico
4 Instituto Venezolano de Investigaciones Científicas (IVIC), Centro de Ingeniería de Materiales y Nanotecnología, Altos de Pipe, Estado Miranda 03940, Venezuela; ericvpp@gmail.com
* Correspondence: ce.alvarado@ugto.mx (C.E.A.-R.); leonardo.sigalotti@gmail.com (L.D.G.S.); Tel.:+52-429-118-3507 (C.E.A.-R.); +52-552-176-1287 (L.D.G.S.)
† All authors contributed equally to this work.

Abstract: A popular equation that is often employed to represent the relationship between the pressure loss and the fluid flow in fluidized or packed granular beds is the Ergun equation, which is an extension of Darcy's law. In this paper, the method of Smoothed Particle Hydrodynamics (SPH) is used to numerically study the flow field across a rectangular channel partially filled with a porous layer both at the Representative Elementary Volume (REV) scale using the Ergun equation and at the pore scale. Since the flow field can be estimated at the REV scale with a much lower cost compared to the pore scale, it is important to evaluate how accurately the pore-scale results can be reproduced at the REV scale. The comparison between both scales is made in terms of the velocity profiles at the outlet of the rectangular channel and the pressure losses across the clear and porous zones for three different arrays of solid grains at the pore scale. The results show that minimum differences in the flow structure and velocity profiles between the REV and the pore scale always occur at intermediate values of the porosity ($\phi = 0.44$ and 0.55). As the porosity increases, the differences between the REV and the pore scale also increase. The details of the pressure losses are affected by the geometry of the porous medium. In particular, we find that the pressure profiles at the REV scale match those at the pore scale almost independently of the porosity only when the grains are uniformly distributed in a non-staggered square array.

Keywords: numerical methods; REV-scale simulation; pore-scale simulation; Ergun equation; packed beds; fluidized beds

1. Introduction

Fluid flow in porous media has been the subject of active research in the last forty years. It happens frequently in geophysical flows as well as in a countless number of engineering applications, such as enhanced oil recovery, carbon dioxide sequestration, water soil infiltration, water filtering, in the design of packed bed reactors, and in many large-scale chemical processes just to mention a few [1–3]. At the pore scale, the movement of solutes in porous materials occurs within an intricate system of interconnected channels or gaps. In highly heterogeneous porous media, the fluid flows through tortuous paths characterized by multiple twists, bends and turns. The combination of Darcy's law and the principle of mass conservation has frequently been employed to explain the fluid flow through a porous medium [4]. The relationship between energy and flow rate is the most important problem in fluid mechanics when dealing with flow through porous media, and Darcy's law is the most basic law that describes this relationship [3]. While the Darcy model is only valid for laminar flows, other semi-empirical models such as the Brinkman–Darcy and the Forchheimer–Darcy models were introduced in the literature. In particular, the Brinkman–Darcy model uses an extra viscous term, called the Brinkman term, which allows the description of the flow through porous media with a porosity higher than 90% [5,6]. On the other hand, the Forchheimer–Darcy model was developed to encompass the regions beyond Darcy flow by incorporating a quadratic velocity term. This term accommodates the non-linear relationship between the pressure gradient and velocity [7,8].

Numerical investigations concerning fluid flow in porous media typically encompass two distinct scales, namely the *pore* scale and the *Representative Elementary Volume* (REV) scale. Pore-scale simulations of porous media are relatively novel. In this scale, the porous medium characteristics are reconstructed in order to obtain representative pore and solid matrix structures in the numerical models [9–16]. All of these numerical investigations were conducted utilizing conventional computational fluid dynamics (CFD) methods, such as finite-difference or finite-volume formulations. In contrast to the pore scale, the REV-scale approach does not require the explicit reconstruction of the porous medium's intricate details. Instead, the presence of the porous medium is accounted for by introducing a resistance term into the momentum equation [17–24]. This translates into a low cost of time compared to most pore-scale simulations by guaranteeing a comparable accuracy. However, most of these simulations have relied on lattice Boltzmann methods to deal with the complex fluid–solid boundaries in the intricate structure of the porous medium. For example, Ergun's equation has recently been applied to model numerically the effects of heat transfer of Cu–water nanofluid through a partially porous wavy channel in the presence of a magnetic field [25] and the magnetohydrodynamical natural convection around a permeable triangular cylinder inside a square enclosure filled with an Al_2O_3–H_2O nanofluid [26].

The flow of fluids through porous media and packed beds of granular solids is subjected to pressure losses due to the simultaneous action of viscous and kinetic energy losses. In the REV-scale approach, the most commonly used equation for calculating the pressure drop has been proposed by Ergun [27]. This equation provides a good basis for estimating the fixed-bed head losses under laminar flow conditions and at rates where the Reynolds number is in the intermediate and turbulent range. It is also quite suitable for determining the flow resistance in packed beds with porosity in the range between 0.4 and 0.6 [28]. Ergun's equation has been extended to describe metal foams, which possess distinct characteristics compared to packed beds, including significantly higher porosity ($\gtrsim$90%) and a unique complex open-celled cellular structure. Specifically, a granular Representative Unit Cell model applicable to all porosities was developed to address these features. For example, du Plessis et al. [29] described the flow of a Newtonian fluid through a packed bed of uniformly sized spherical granules. Moreover, Bazmi et al. [30] modified the Ergun equation for predicting the pressure losses in trickle bed reactors randomly packed with trilobe particles. More recently, a modification of the standard Ergun equation by multiplying it with a resistance correction coefficient was proposed by Lai et al. [31] to calculate head losses through a high-porosity, open-cell structured porous medium in a rectangular channel. In a more recent study conducted by Pang et al. [32], the intrinsic permeability of crushed coal medium was experimentally determined using the Ergun equation. This research offers valuable insights into the permeability characteristics of the media within fractured coal bodies. An evaluation of the pressure drop in heterogeneous alternating-layer packed beds was modeled by Li et al. [33] using a coupled LES-LBM-IMB-DEM method to obtain the interphase closure of fluid-solid systems. A discrete element method was also applied by Li et al. [34] to evaluate pressure losses in heterogeneous alternating-layer beds for low-carbon operating blast furnaces.

The aforementioned numerical investigations were carried out utilizing conventional computational fluid dynamics (CFD) schemes, including finite difference, finite volume, and lattice Boltzmann formulations. However, in this paper, the authors employ the Smoothed Particle Hydrodynamics (SPH) method, which was initially developed by Lucy [35] and Gingold and Monaghan [36] for the simulation of astrophysical flows, to study the fluid flow through porous media in the REV scale. The SPH method is a Lagrangian scheme that relies on particle interpolation to calculate smooth field variables [37,38]. Since its inception, SPH has become very popular because of its ever-increasing applications to fluid and solid mechanics, among many other fields of science and engineering [39]. Although the SPH method is relatively novel compared to more traditional mesh-based schemes, it is not new to flow modeling in porous media [40,41]. However, most SPH simulations of flow through porous media have been designed using the pore-scale approach [42–47]. For example, Jiang et al. [42] employed SPH techniques to simulate fluid flow in isotropic porous media. These authors employed SPH methods and resolved the porous structure at a mesoscopic level by randomly assigning a portion of the SPH particles to fixed locations, while Tartakovsky et al. [43] simulated numerically the fluid flow through anisotropic and heterogeneous porous media. Multiphase flows in porous media with SPH were also modeled by Tartakovsky et al. [44], Kunz et al. [45] and Peng et al. [47] using the pore-scale approach. SPH models of fluid flowing through a deformable porous medium were reported by Lenaerts et al. [48] and Bui and Nguyen [49]. These researchers conducted simulations of porous flow on a macroscopic scale by abstracting individual pores or cavities within the material. On the other hand, the scheme introduced by the latter authors is based on Biot's two-phase mixture theory. In this scheme, the solid matrix is represented as an elasto-plastic material, while the pore fluid is modeled as an incompressible fluid. SPH simulations of fluid flow through fractured porous media were also recently reported by Shigorina et al. [50] and by Bui and Nguyen [51].

As far as we know, there are no reports in the literature on the SPH modeling of flow in porous media at the REV scale. Since SPH is a Lagrangian scheme, it presents some advantages over traditional Eulerian mesh-based methods in that fluid advection is performed exactly. On the other hand, including more physics to the model is a straightforward task compared to other schemes. Although the pore-scale approach appears to be more accurate and allows studying microscopic seepage and fluid–solid interactions, for large porous systems, it involves a much higher computational cost. Therefore, exploring the ability of SPH to model the flow in porous media at the REV scale with an accuracy comparable to the pore-scale approach will be a valuable task. The performance of SPH is tested for the same benchmark problem introduced by Lai et al. [31], namely flow through a rectangular channel that includes a porous medium at different porosities. In order to measure the accuracy of the simulations at the REV scale a comparison is made with identical models at the pore scale. The paper is organized as follows. A brief description of the governing equations and SPH formalism is given in Section 2. Section 3 deals with the problem statement and boundary conditions, while the validation and convergence tests are presented in Section 4. Section 5 describes the results and Section 6 contains the main conclusions.

2. Governing Equations and SPH Formulation

2.1. Basic Equations

In Eulerian form, the differential equations describing the laminar flow through a homogeneous porous medium at the REV scale are given by the macroscopic continuity and momentum equations [52,53]:

$$\frac{\partial \rho}{\partial t} + \nabla \cdot (\rho \mathbf{v}) = 0, \tag{1}$$

and

$$\frac{\partial (\rho \mathbf{v})}{\partial t} + \nabla \cdot \left(\frac{\rho \mathbf{v}\mathbf{v}}{\phi} \right) = -\nabla p + \nabla \cdot \left[\rho \nu (\nabla \mathbf{v} + \nabla \mathbf{v}^t) \right] + \mathbf{F}, \tag{2}$$

respectively, where ρ is the fluid mass density, $\mathbf{v}$ is the fluid velocity vector, p is the pressure, ν is the kinematic viscosity, ϕ is the porosity of the porous medium (i.e., the ratio between the void volume and the total volume), $\mathbf{F}$ is the resistance force term (or Brinkman term), which will be described below, and the superscript t means transposition. The form of the momentum equation is derived from first principles over a fluid saturated porous control volume [52].

Since SPH is a fully Lagrangian scheme, Equations (1) and (2) must be written using Lagrangian coordinates. After simple algebraic steps, this set of equations becomes

$$\frac{d\rho}{dt} = -\rho\nabla\cdot\mathbf{v}, \tag{3}$$

$$\begin{aligned}\frac{d\mathbf{v}}{dt} &= -\frac{1-\phi}{\phi}\left[\mathbf{v}\nabla\cdot\mathbf{v}+\mathbf{v}\cdot\nabla\mathbf{v}+\frac{1}{\rho}\mathbf{v}\mathbf{v}\cdot\nabla\rho\right]- \\ &- \frac{1}{\rho}\nabla p+\frac{1}{\rho}\nabla\cdot\left[\rho\nu(\nabla\mathbf{v}+\nabla\mathbf{v}^t)\right]+\frac{1}{\rho}\mathbf{F},\end{aligned} \tag{4}$$

where d/dt is the material time derivative. In the derivation of Equation (4), the assumption was made that $\nabla\phi = 0$. The dynamical pressure is related to the density by means of the Murnaghan–Tait equation of state [54]:

$$p = p_0\left[\left(\frac{\rho}{\rho_0}\right)^\gamma - 1\right], \tag{5}$$

where $\gamma = 7$, $p_0 = c_0^2\rho_0/\gamma$, ρ_0 is a reference density and c_0 is a numerical sound speed. In order to ensure weak compressibility and fluctuations of the density field $|\rho-\rho_0|/\rho_0 \lesssim 0.01$, the value of c_0 is taken to be at least 10 times higher than the maximum fluid velocity over the entire system [54].

2.2. Ergun Equation

The term $\mathbf{F}$ in Equations (2) and (4) is the total body force due to the presence of a porous medium. This term is given by the classical Ergun equation [27]:

$$\mathbf{F} = -\frac{\phi\nu}{K}\mathbf{v} - \frac{\phi F_\phi}{\sqrt{K}}|\mathbf{v}|\mathbf{v}, \tag{6}$$

where K is the permeability of the porous medium, which is given by

$$K = \frac{\phi^3 d_p^2}{150(1-\phi)^2}, \tag{7}$$

and F_ϕ is the structure (o geometric) function, which is defined by the relation

$$F_\phi = \frac{1.75}{\sqrt{150\phi^3}}. \tag{8}$$

The factor d_p in the numerator of Equation (7) is the mean diameter of the solid particles forming the structure of the porous medium and is calculated as

$$d_p = \frac{2D(1-\phi)}{S}, \tag{9}$$

where D is the dimension ($D = 2$ in two dimensions and $D = 3$ in three dimensions) and S is the specific surface area of the porous medium, which is defined as the ratio of the surface area of grains to the volume of rock grains [55]. Here, the specific surface area of the porous medium is estimated to be $S = 1.35\langle P_s\rangle$ [56], where $\langle P_s\rangle$ is the averaged ratio of the grains perimeter to their area in the two-dimensional images. In this work, the rock grains are considered to be perfectly spherical and therefore $S = 3\langle P_s\rangle/2$. The values of $\langle P_s\rangle$ for the different porous layer configurations are calculated following the algorithm described by Rabbani et al. [56]. Note that the definition given in Equation (7) for the permeability is just the Darcy number, Da. As it stands, Equation (6) holds for the full range of laminar, transitional and turbulent flow through a uniform fixed bed. The porosity ϕ is a number varying in the interval $0 \leq \phi \leq 1$. When $\phi = 0$, there are no void spaces, implying a null porosity, while if $\phi = 1$, there is no porous medium. In this case, $\mathbf{F} = \mathbf{0}$ and Equation (4) reduces to the pore-scale equation.

2.3. Large-Eddy Simulation (LES) Filtering

A LES filtering is applied to Equations (3) and (4). In this approach, the fluid velocity is separated into two terms, namely the mean velocity component (or resolved scale velocity), $\tilde{\mathbf{v}}$, and its fluctuating part (or sub-particle scale velocity), $\mathbf{v}'$, such that $\mathbf{v} = \tilde{\mathbf{v}} + \mathbf{v}'$. The mean velocity component is obtained by a density-weighted Favre filtering, i.e.,

$$\tilde{\mathbf{v}} = \frac{1}{\bar{\rho}} \frac{1}{T} \int_t^{t+T} \rho(\mathbf{x}, t)\mathbf{v}(\mathbf{x}, t)dt, \tag{10}$$

where $\mathbf{x} = (x, y, z)$, T is a time interval and $\bar{\rho}$ is the mean density given by the conventional Reynolds-averaged density. Applying the Favre filtering to Equations (3) and (4) yields the spatially filtered equations

$$\frac{d\bar{\rho}}{dt} = -\bar{\rho}\nabla \cdot \tilde{\mathbf{v}}, \tag{11}$$

$$\begin{aligned} \frac{d\tilde{\mathbf{v}}}{dt} &= -\frac{1-\phi}{\phi}\left[\tilde{\mathbf{v}}\nabla \cdot \tilde{\mathbf{v}} + \tilde{\mathbf{v}} \cdot \nabla\tilde{\mathbf{v}} + \frac{1}{\bar{\rho}}\tilde{\mathbf{v}}\tilde{\mathbf{v}} \cdot \nabla\bar{\rho}\right] - \\ &- \frac{1}{\bar{\rho}}\nabla\bar{p} + \frac{\nu}{\bar{\rho}}[\nabla \cdot (\bar{\rho}\nabla)]\tilde{\mathbf{v}} + \frac{\nu}{\bar{\rho}}\nabla \cdot \tilde{\mathbf{T}} + \frac{1}{\bar{\rho}}\tilde{\mathbf{F}}, \end{aligned} \tag{12}$$

where $\tilde{\mathbf{F}}$ obeys the same expression (6), but with $\mathbf{v} \to \tilde{\mathbf{v}}$, and $\tilde{\mathbf{T}}$ is the sub-particle stress tensor, which in index notation has the form

$$\tilde{T}_{ij} = \bar{\rho}\nu_t\left(2\tilde{S}_{ij} - \frac{2}{3}\tilde{S}_{kk}\delta_{ij}\right) - \frac{2}{3}\bar{\rho}C_I\nabla^2\delta_{ij}|\tilde{S}|^2, \tag{13}$$

where $\tilde{S}_{ij}$ is the Favre-filtered strain tensor given by

$$\tilde{S}_{ij} = \frac{1}{2}\left(\frac{\partial\tilde{v}_i}{\partial x_j} + \frac{\partial\tilde{v}_j}{\partial x_i}\right), \tag{14}$$

$C_I = 0.00066$, $\nu_t = (0.12\nabla)^2|\tilde{S}|$ is the Smagorinsky eddy viscosity, $|\tilde{S}| = (2\tilde{S}_{ij}\tilde{S}_{ij})^{1/2}$ is the local strain rate, δ_{ij} is the Kronecker delta and Δ is a measure of the finite particle size. For practical purposes, Δ is set equal to the smoothing length h (see below).

2.4. SPH Solver

Equations (11) and (12) are solved in two-space dimensions using the last updated version of the DualSPHysics code [57], which relies on SPH methods. The SPH discretization of the governing equations is obtained by dividing the model domain into N subdomains, each of which contains a Lagrangian particle [37]. In essence, the SPH representation of a function is based on two approximations, namely the kernel and the particle approximation. In the kernel approximation, a smooth function, $f = f(\mathbf{x})$, is estimated by convolving the function itself with an interpolating kernel, such that

$$\langle f(\mathbf{x})\rangle = \int_\Omega f(\mathbf{x}')W(|\mathbf{x} - \mathbf{x}'|, h)d\mathbf{x}', \tag{15}$$

where $W(|\mathbf{x}-\mathbf{x}'|, h)$ is the kernel function, h is the width of the kernel (or smoothing length) and the integration is taken over the entire problem domain Ω.

A particle estimate of the function is obtained by replacing the integral in Equation (15) by a Riemann sum over the compact support of the kernel, i.e., over the region $|\mathbf{x}-\mathbf{x}'| \leq kh$, where $W \neq 0$. This way, for each particle a, the SPH value of the function is given by the expression

$$f_a = \sum_{b=1}^{n} \frac{m_b}{\rho_b} f_b W_{ab}, \tag{16}$$

where $W_{ab} = W(|\mathbf{x}_a - \mathbf{x}_b|, h)$, and n is the number of neighbors of the observation particle a falling within its kernel support. In DualSPHysics, Equation (11) is convolved and discretized using the approach proposed by Monaghan [37], namely

$$\frac{d\rho_a}{dt} = -\rho_a \sum_{b=1}^{n} \frac{m_b}{\rho_b} (\mathbf{v}_a - \mathbf{v}_b) \cdot \nabla_a W_{ab}, \tag{17}$$

where the tilde and bar operators over the ensemble average velocity vector and mean density, respectively, are dropped for simplicity. The SPH representation of Equation (12) is

$$\begin{aligned}
\frac{d\mathbf{v}_a}{dt} = & - \frac{1-\phi}{\phi} \sum_{b=1}^{n} \frac{m_b}{\rho_b} [\mathbf{v}_a(\mathbf{v}_b - \mathbf{v}_a) \cdot \nabla_a W_{ab} + \mathbf{v}_a \cdot (\mathbf{v}_b - \mathbf{v}_a) \nabla_a W_{ab}] \\
& - \frac{1-\phi}{\phi} \sum_{b=1}^{n} \frac{m_b}{\rho_a \rho_b} (\rho_b - \rho_a) \mathbf{v}_a \mathbf{v}_a \cdot \nabla_a W_{ab} \\
& - \frac{1}{\rho_a} \sum_{b=1}^{n} \frac{m_b}{\rho_b} (p_a + p_b) \nabla_a W_{ab} + 4\nu \sum_{b=1}^{n} m_b \frac{\mathbf{v}_a - \mathbf{v}_b}{\rho_a + \rho_b} \frac{(\mathbf{x}_a - \mathbf{x}_b) \cdot \nabla_a W_{ab}}{|\mathbf{x}_a - \mathbf{x}_b|^2 + \epsilon^2} \\
& + \sum_{b=1}^{n} m_b \left(\frac{\mathbf{T}_a}{\rho_a^2} + \frac{\mathbf{T}_b}{\rho_b^2} \right) \cdot \nabla_a W_{ab} + \mathbf{F}_a,
\end{aligned} \tag{18}$$

where $\epsilon = 0.1h$ whenever the distance between particles a and b happens to be close to zero. The symmetric representation proposed by Colagrossi and Landrini [58] is employed for the pressure gradient on the basis that this form is variationally consistent with the representation (17) of the continuity equation [59]. Moreover, the laminar viscous term and the sub-particle stress term are discretized using the replacements suggested by Lo and Shao [60]. In order to prevent amplification of numerical errors due to anisotropies in the particle distribution, the SPH particles are moved using the equation

$$\frac{d\mathbf{x}_a}{dt} = \mathbf{v}_a + \frac{\beta x_0 v_{\max}}{M} \sum_{b=1}^{N} m_b \frac{\mathbf{x}_a - \mathbf{x}_b}{|\mathbf{x}_a - \mathbf{x}_b|^3}, \tag{19}$$

where $\beta = 0.04$, $v_{\max}$ is the maximum estimated fluid velocity, M is the total fluid mass and

$$x_0 = \frac{1}{N} \sum_{b=1}^{N} |\mathbf{x}_a - \mathbf{x}_b|, \tag{20}$$

where now the summations in Equations (19) and (20) are taken over all particles of the system. To improve the convergence properties of SPH, a Wendland C^2 function [61]

$$W(q,h) = \alpha_D \left(1 - \frac{q}{2}\right)^4 (2q+1) \quad \text{for} \quad 0 \leq q \leq 2, \tag{21}$$

is employed, where $\alpha_D = 7/(4\pi h^2)$ in two dimensions and $21/(16\pi h^3)$ in three dimensions and $q = |\mathbf{x}-\mathbf{x}'|/h$. The time integration of Equations (17)–(19) is performed using the symplectic integrator provided by DualSPHysics, which ensures second-order accuracy and numerical coupling of the SPH equations when the density, velocity and position of particles are advanced from time t^n to $t^{n+1} = t^n + \Delta t$.

3. Model Problem and Boundary Conditions

The effects of a porous substrate on the incompressible, laminar flow of water in a rectangular channel of length $L = 0.1$ m and width $H = 0.05$ m are investigated in two-space dimensions. As shown in Figure 1, the channel is subdivided into a clear zone and a rectangular porous layer in its central core of length $L_{pm} = 0.05$ m and width $H_{pm} = 0.025$ m. The flow is assumed to be isothermal and fully developed so that it enters the channel with a prescribed parabolic velocity profile (see Figure 1). All numerical experiments were conducted using water at room temperature as the working fluid ($\rho = 1000$ kg m^{-3}). As displayed in the bottom right picture of Figure 1, the maximum velocity at the inlet is $v_0 = 1.25 \times 10^{-3}$ m s^{-1}, corresponding to a Reynolds number, Re $= 0.0125$. This model problem is the same as that employed by Lai et al. [31] in their lattice Boltzmann calculations with a corrected Ergun equation for high porosity and open-celled substrates. In contrast to Lai et al. [31], here, the simulations were conducted for substrate porosities $\phi = 0.3, 0.44, 0.55, 0.77$ and 0.88. The simulations at the REV scale were performed using the same geometry and parameters as the simulations at the pore scale. In this latter case, three different porous geometries were studied, corresponding to regular non-staggered square, regular staggered square and random arrays of circular grains, as shown in the frames of the left column of Figure 1 for the particular case when $\phi = 0.44$.

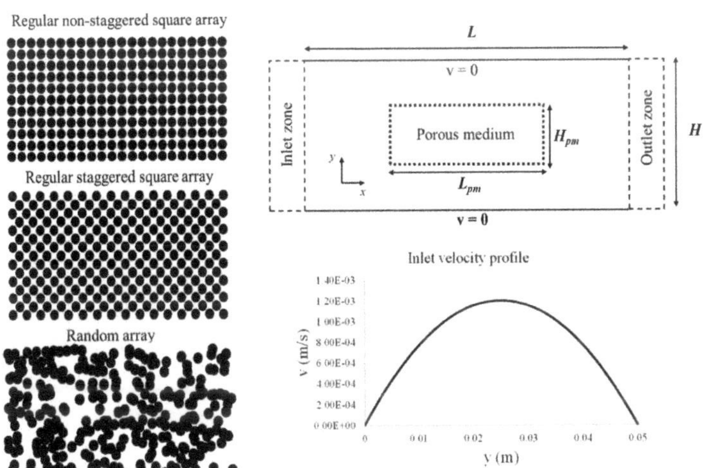

Figure 1. Geometries of the porous media employed in the pore-scale simulations for the case when $\phi = 0.44$ (**left column of frames**), schematic diagram of the problem at REV scale (**top right**) and steady-state inlet velocity profile used in all simulations (**bottom right**).

No-slip boundary conditions are applied at the walls of the channel ($\mathbf{v} = \mathbf{0}$), using the method of dynamic boundary particles implemented in DualSPHysics [57]. The implementation of this boundary condition parallels that described by Alvarado-Rodríguez et al. [62] for flow in rigid pipes of circular cross-section. Non-reflective outflow boundary conditions are implemented at the exit of the channel [63]. This method allows for anisotropic wave propagation across the outlet plane, where the velocity vector of particles in the outflow zone is evolved using the outgoing wave equation

$$\frac{\partial \mathbf{v}}{\partial t} + v_x \frac{\partial \mathbf{v}}{\partial x} - \nu \frac{\partial^2 \mathbf{v}}{\partial y^2} = \mathbf{0}, \tag{22}$$

where $\mathbf{v} = (v_x, v_y)$. Here, the v_x-velocity component corresponds to the mainstream velocity along the channel. For an outflow particle, say o, the above equation admits the numerically stable SPH representation

$$\frac{\partial \mathbf{v}_o}{\partial t} = -v_{x,o} \sum_{b=1}^{n} \frac{m_b}{\bar{\rho}_{ob}} (\mathbf{v}_b - \mathbf{v}_o) \frac{\partial W_{ab}}{\partial x_o} + 2\nu \sum_{b=1}^{n} \frac{m_b}{\rho_b} \frac{y_{ob}(\mathbf{v}_b - \mathbf{v}_o)}{|\mathbf{x}_{ob}|^2 + \epsilon^2} \frac{\partial W_{ab}}{\partial y_o}, \tag{23}$$

where $\mathbf{x}_{ob} = \mathbf{x}_o - \mathbf{x}_b$, $y_{ob} = y_o - y_b$ and $\bar{\rho}_{ob} = (\rho_o + \rho_b)/2$. According to Equation (23), outflow particles near the outlet plane will have some neighbors pertaining to the fluid domain behind the outlet, thus allowing fluid information to be conveyed into the outflow zone with no noise reflection into the fluid domain. The position of outflow particles is evolved by solving the equation

$$\frac{d\mathbf{x}_o}{dt} = \mathbf{v}_o, \tag{24}$$

which is then time-integrated simultaneously with Equation (23) using the same symplectic integrator employed for the hydrodynamics.

4. Validation and Convergence Testing

The Dualphysics code has been thoroughly validated against experimental data in pipe flow simulations using different geometries [62,64] and for many other benchmark test cases [57]. As a validation test, we further consider unsteady plane Poiseuille flow in a rectangular channel, while the convergence test is performed by comparing the REV-scale mainstream velocity profile at the outlet plane of the rectangular channel with that obtained at the pore scale for a uniform, non-staggered array of grains for $\phi = 0.3$ at varying spatial resolution.

As shown in Figure 2, the channel consists of two infinite, parallel plates, and the (x, y)-plane is chosen to represent the flow with the positive x-axis as the flow direction. The plates are located at distances $H_0/2 = \pm 0.0005$ m from the center ($y = 0$) of the coordinate system. For this problem, the y-components of the velocity and pressure gradient vanish identically so that the x-component of the velocity depends only on y and time. The particles are initially at rest and moved through the channel by applying a body force, F_b, due to the hydrostatic pressure gradient $\Delta p/L_0$

$$F_b = \frac{1}{\rho}\left(\frac{\Delta p}{L_0}\right) = -\frac{8\nu v_0}{H_0^2}, \tag{25}$$

where $L_0 = 0.005$ m is the length of the channel plates and v_0 is the velocity at the vertex of the asymptotic (steady-state) velocity profile

$$v_x(y) = \frac{F_b}{2\nu}\left(y^2 - \frac{H_0^2}{4}\right) = v_0\left(1 - \frac{4y^2}{H_0^2}\right). \tag{26}$$

We choose initial parameters similar to those employed by Morris et al. [65] and Sigalotti et al. [66], i.e., $\rho_0 = 1000$ kg m^{-3}, $v_0 = 1.25 \times 10^{-5}$ m s^{-1} and Re $= 0.0125$, resulting in a kinematic viscosity $\nu = H_0 v_0/\mathrm{Re} = 1.0 \times 10^{-6}$ m^2 s^{-1}.

Convergence to the analytical solution was assessed by increasing the spatial resolution from $N = 550$ to 50,601 SPH fluid particles. For simplicity, the particles are initially arranged in a regular Cartesian mesh, and the simulations are followed until a steady-state solution is achieved. For the above resolutions, the initial interparticle distance varies between 0.1 mm (for $N = 550$) and 0.01 mm (for N = 50,601). Inlet and outlet boundary conditions during the calculations were handled by means of a cyclic boundary condition, which is enforced by placing five columns of imaginary particles to the left ($y < 0$) and to the right ($y > L_0$) of the fluid domain. The right picture of Figure 2 shows the numerically obtained profiles (symbols) as compared with the analytical solution (solid line) at $t = 0.6$ s, when the simulations have reached a steady-state solution. At the maximum resolution tried, the numerical profile (filled dots) closely overlaps the analytical solution with a deviation in terms of a root-mean-square error (RMSE) $\lesssim 5.3 \times 10^{-6}\%$. At this resolution, the numerical results show a very good matching with the exact solution.

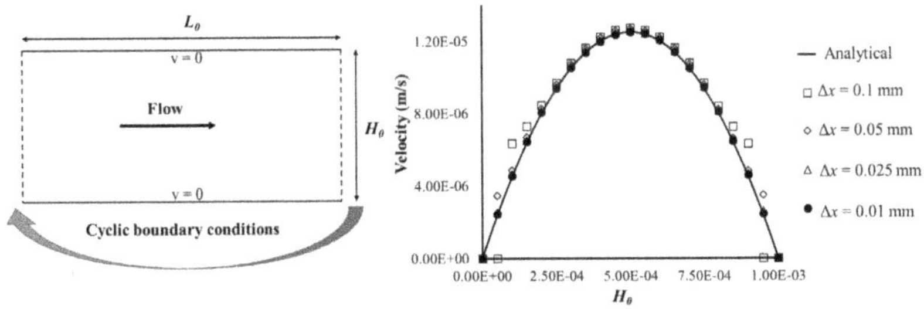

Figure 2. Schematic diagram showing the geometry for the plane Poiseuille flow test problem (**left picture**) and numerical steady-state velocity profiles at different spatial resolutions (symbols) as compared to the analytical solution (solid line).

In order to provide internally coherent flow conditions for the subsequent simulations of flow through a rectangular channel with a partially filled porous layer, a convergence test was carried out by varying the total number of particles, N. Upon increasing the number of particles, the results at the REV scale are compared with those obtained at the pore scale using 8,006,801 SPH particles for the case when the porous layer consists of a uniform, non-staggered array of grains with porosity $\phi = 0.3$. This value is below the range $0.4 \leq \phi \leq 0.6$ for which the Ergun equation is customarily applied to calculate the flow resistance of packed beds and spherical granular materials [28]. Convergence to the pore-scale solution is measured by varying the number of particles at the REV scale from N = 1,082,952 to N = 8,006,801. The results of the convergence study are reported in Figure 3. As shown in Figure 3a, the mainstream velocity profiles at the REV and pore scale are compared at the exit plane of the rectangular channel at different spatial resolutions from N = 1,082,952 ($\Delta x = 0.068$ mm) to N = 8,006,801 ($\Delta x = 0.025$ mm), while the same is displayed in Figure 3b for the mainstream velocity profile along the channel in the top clear zone. The deviations between both profiles in terms of the RMSE metric are: $\sim 7.42 \times 10^{-5}$ m s^{-1} (for N = 1,082,952), $\sim 4.43 \times 10^{-5}$ m s^{-1} (for N = 2,003,401), $\sim 2.20 \times 10^{-5}$ m s^{-1} (for N = 4,084,654) and $\sim 1.98 \times 10^{-5}$ m s^{-1} (for N = 8,006,801) in Figure 3a, while in Figure 3b, the RMSEs are: $\sim 1.58 \times 10^{-4}$ m s^{-1} (for N = 1,082,952), $\sim 9.93 \times 10^{-5}$ m s^{-1} (for N = 2,003,401), $\sim 4.66 \times 10^{-5}$ m s^{-1} (for N = 4,084,654) and $\sim 1.89 \times 10^{-5}$ m s^{-1} (for N = 8,006,801). In both cases, the numerical profiles at the REV scale show an asymptotic tendency to globally converge to the highly resolved pore-scale solution as the number of particles is increased. The pore-scale simulations use exactly the same SPH solver and also the same number of SPH particles. In both cases, the actual number of particles (i.e., N = 8,006,801) was chosen based on convergence considerations.

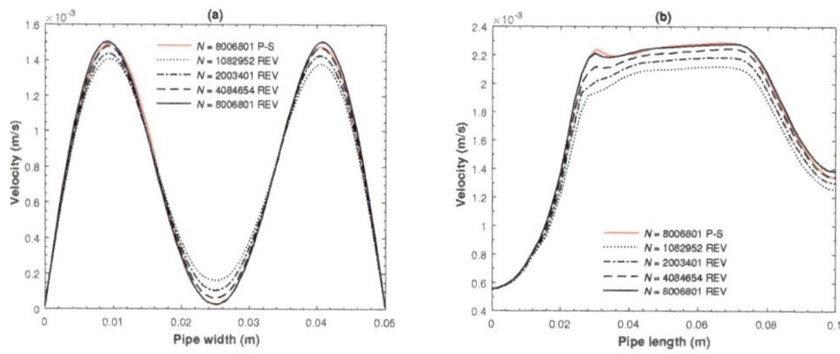

Figure 3. (**a**) REV-scale velocity profiles at the exit plane of the rectangular channel at different spatial resolutions as compared with the pore-scale solution with $N = 8,006,801$ SPH particles (red curve). (**b**) The same for the mainstream velocity profile along the channel in the top clear zone.

5. Results

The results from the REV-scale simulations are compared with those obtained at the pore-scale approach for the three different grain geometries shown in Figure 1. The comparison here is enforced in terms of the mean velocity profiles at the outlet section of the rectangular channel and the pressure losses across the porous medium.

5.1. Flow Structure and Velocity Profiles

A comparison of the mainstream velocity field across the channel between the pore scale (for the three grain geometries) and the REV scale is shown in Figure 4 for the case when the porosity is $\phi = 0.44$. Similar plots are also seen for the other porosity values. In the four cases, the flow structure looks qualitatively similar in the clear zone and across the porous layer. The flow similarity between the pore and REV scales is not surprising, because the classical Ergun equation is known to work satisfactorily well for intermediate porosity values in the range $0.4 \lesssim \phi \lesssim 0.6$ [28]. However, quantitative differences may still exist. For example, the maximum velocity in the clear zone is somewhat slower at the REV scale $v_{max} = 1.4185 \times 10^{-3}$ m s^{-1} (Figure 4d) compared to the pore-scale simulations $v_{max} = 1.4611 \times 10^{-3}$ m s^{-1}, $v_{max} = 1.4923 \times 10^{-3}$ m s^{-1} and $v_{max} = 1.44451 \times 10^{-3}$ m s^{-1} (Figures 4a–c), respectively, where the maximum velocity increases as the geometry of the packed column is changed from a non-staggered to a staggered square to a completely random array of grains.

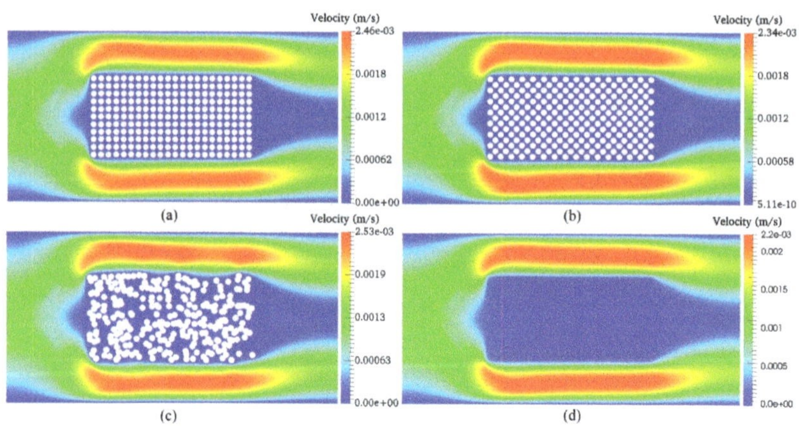

Figure 4. Mainstream fluid velocity across the channel obtained from the pore-scale simulations when the interposed porous layer consists of (**a**) a non-staggered square, (**b**) a staggered square and (**c**) a random array of circular grains as compared to (**d**) the REV-scale simulation for $\phi = 0.44$.

The resulting velocity profiles at the outlet plane ($x = L = 0.1$ m) of the channel are displayed in Figure 5 for the REV- and the three different pore-scale simulations when the porosity is varied from $\phi = 0.44$ to 0.88. In all cases, the velocity profiles exhibit a wave-like form, where the crests correspond to the maximum flow velocity of fluid coming from the clear zones of the channel (dim yellow streams at the right border planes of all frames in Figure 4) and the valley to the minimum velocity of the fluid coming from the central porous layer (center of blue stripes at the right border planes of all frames in Figure 4).

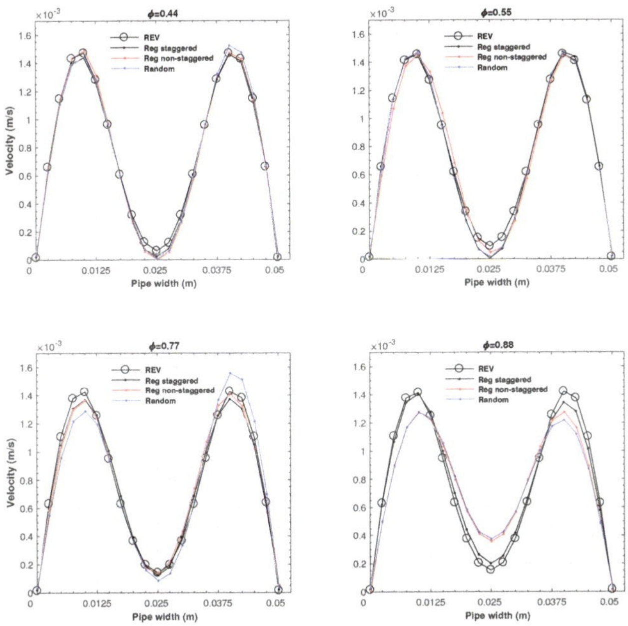

Figure 5. Mainstream velocity profiles at the outlet plane of the channel for the REV- and pore-scale simulations at different porosity values between $\phi = 0.44$ and 0.88. The profiles obtained at the pore scale for the three different geometries of the porous layer are compared with the REV-scale results.

For $\phi = 0.3$, there is little difference between the REV and the pore-scale profiles when the packed columns consist of non-staggered and staggered uniformly distributed grains. However, at such low porosity, the difference between the REV- and the pore-scale profiles are more pronounced around the zones of maximum velocities (crests) when the porous layer consists of randomly distributed grains. At a higher porosity, corresponding to $\phi = 0.44$ and 0.55, the results for the REV-scale and the pore-scale simulations (for all three packed geometries) exhibit very small differences, with the minimum velocities around the valley being slightly overestimated by the REV-scale approach. When the porosity grows to $\phi = 0.77$ the REV-scale profiles differ from the pore-scale ones mostly for the case of randomly packed grains. At even higher porosity values, as is the case for $\phi = 0.88$, the differences between the REV- and the pore-scale simulations become even larger independently of the geometry of the porous layer.

Figure 6 shows the root-mean-square errors (RMSEs) between the REV-scale and the pore-scale velocity profiles of Figure 5. For each porosity value, the distances between the REV-scale profile and the pore-scale results are displayed for all three geometries of the porous layer. In general, the error is seen to increase with the porosity. It is interesting to see that at porosity values of 0.44 and 0.55, the errors between the REV- and the pore-scale results are almost independent of the geometry of the porous layer. However, at higher porosity, the errors depend more strongly on the geometry of the porous layer and become larger for randomly distributed grains. A similar trend was also observed for $\phi = 0.3$.

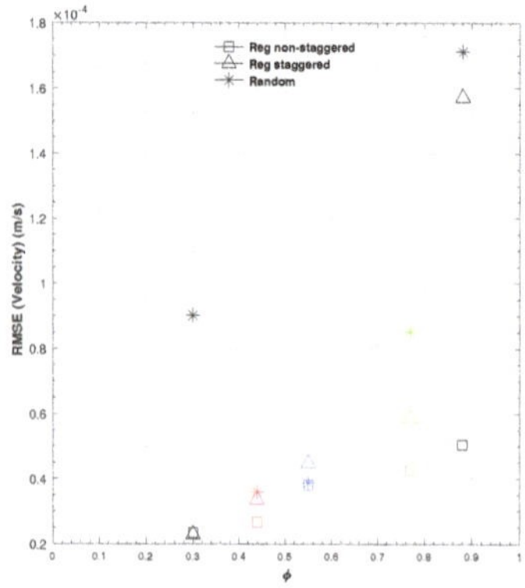

Figure 6. Root-mean-square error (RMSE) as a function of the porosity for the velocity profiles at the outlet plane of the channel (see Figure 5). The symbols identify the geometry of the porous layer.

5.2. *Pressure Losses*

Figure 7 displays the pressure losses along the rectangular channel in the clear zones (top and bottom) and center of the porous layer at different porosity values. A comparison between the REV-scale (blue curves) and the pore-scale simulations (black curves) for the case when the grains are distributed in a non-staggered regular array is presented. The results of the REV-scale calculation closely follow those predicted by the pore-scale simulation when $\phi = 0.44$ and 0.55, while the distance between both models progressively increases at higher porosity values. However, the RMSEs between the REV- and the pore-scale results remain sufficiently small even in the worst case when $\phi = 0.88$ (see Figure 7). In this work, we are interested in demonstrating that Ergun model works well for intermediate and relatively high porosity values $0.44 < \phi < 0.88$. Moreover, the trends of the pressure along the channel were found to be very similar to those displayed at higher porosities, although the distance between the REV scale and any of the pore-scale geometries reported was a bit longer at $\phi = 0.3$.

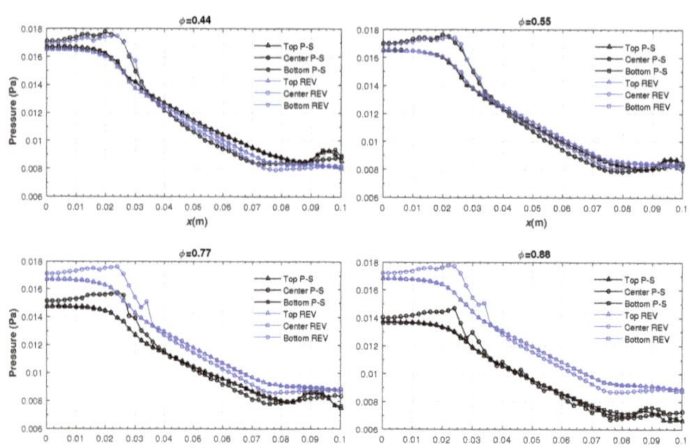

Figure 7. Pressure losses predicted by the REV-scale simulation (blue curves) as compared with the pore-scale calculation when the porous layer consists of grains distributed in a regular non-staggered way (black curves). The results are displayed for the top and bottom clear zones and for the center of the porous layer.

An inspection of Figure 7 shows that the flow through the porous layer experiences mean pressure drops of $\approx 8.3 \times 10^{-3}$ Pa for $\phi = 0.44$, $\approx 8.8 \times 10^{-3}$ Pa for $\phi = 0.55$, $\approx 6.7 \times 10^{-3}$ Pa for $\phi = 0.77$ and $\approx 6.7 \times 10^{-3}$ Pa for $\phi = 0.88$ at the pore scale. The pressure drop increases as the porosity decreases. This result is consistent with what is expected if the properties of the fluid, such as surface tension, contact angle and density, remain constant. Moreover, under a greater number of obstacles in the porous layer, the sizes of the porous throats decrease, which produces an increase in the flow velocity there and a corresponding greater variation in the pressure.

The results when the geometry of the porous layer consists of a staggered regular distribution of grains are displayed in Figure 8. It is clear that the details of the pressure losses are sensitive to the geometry of the porous medium. In contrast to Figure 7, in this case, the results at the REV scale do not match the ones at the pore scale for porosity values $\phi \leq 0.77$. In all cases, the REV-scale simulations predict pressure losses greater than the pore-scale calculations through the clear zones and the porous layer. Only when $\phi = 0.88$ do the pressure losses appear to follow similar trends for both models. From Figure 8, it follows that compared to Figure 7, the flow through the porous layer undergoes mean pressure losses of $\approx 5.3 \times 10^{-3}$ Pa for $\phi = 0.44$, $\approx 5.1 \times 10^{-3}$ Pa for $\phi = 0.55$, $\approx 4.4 \times 10^{-3}$ Pa for $\phi = 0.77$ and $\approx 6.3 \times 10^{-3}$ Pa for $\phi = 0.88$. Figure 9 shows the pressure drops when the geometry of the porous layer consists of randomly distributed grains. The trends are qualitatively similar to the variations displayed in Figure 8. In no case are the profiles at the REV scale matching those at the pore scale when the grains are randomly distributed, implying that the details of the pressure profiles are sensitive to the geometry of the porous medium. In this case, the pressure drops through the porous layer are $\approx 5.9 \times 10^{-3}$ Pa for $\phi = 0.44$, $\approx 5.5 \times 10^{-3}$ Pa for $\phi = 0.55$, $\approx 4.7 \times 10^{-3}$ Pa for $\phi = 0.77$ and $\approx 3.3 \times 10^{-3}$ Pa for $\phi = 0.88$. In this case, a clear tendency exists for the pressure drop to increase for decreasing porosity. For comparison, the mean pressure losses at the REV scale (blue curves in Figures 7–9) are: $\approx 8.6 \times 10^{-3}$ Pa for $\phi = 0.44$, $\approx 8.4 \times 10^{-3}$ Pa for $\phi = 0.55$, $\approx 8.3 \times 10^{-3}$ Pa for $\phi = 0.77$ and $\approx 8.3 \times 10^{-3}$ Pa for $\phi = 0.88$. Evidently, these figures show that the mean pressure losses across the porous layer are overestimated at the REV scale, which predict larger pressure drops than the pore-scale simulations.

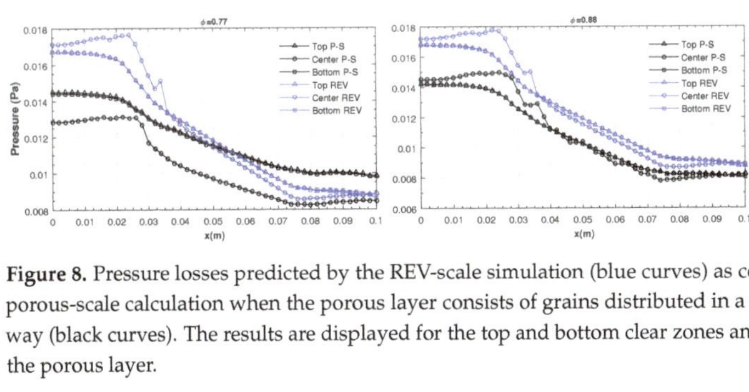

Figure 8. Pressure losses predicted by the REV-scale simulation (blue curves) as compared with the porous-scale calculation when the porous layer consists of grains distributed in a regular staggered way (black curves). The results are displayed for the top and bottom clear zones and for the center of the porous layer.

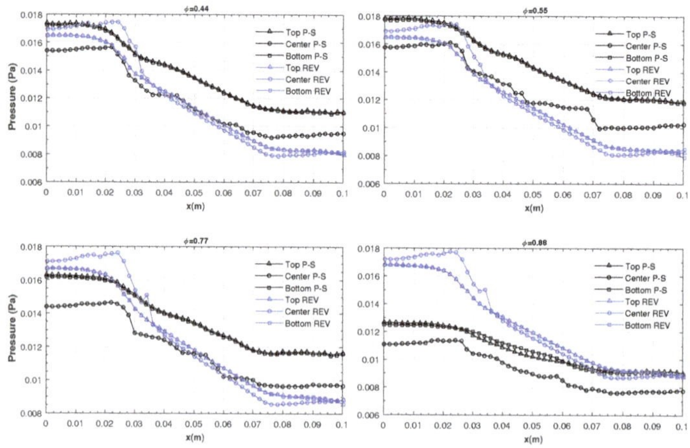

Figure 9. Pressure losses predicted by the REV-scale simulation (blue curves) as compared with the porous-scale calculation when the porous layer consists of randomly distributed grains (black curves). The results are displayed for the top and bottom clear zones and for the center of the porous layer.

However, as depicted in Figure 10a–c, the RMSEs between the REV- and the pore-scale pressure profiles remain sufficiently small independently of the geometry of the porous layer. However, the actual distance between the REV- and the pore-scale results increases at higher porosity values. For the case of a uniform non-staggered array of grains, the smallest RMSEs between the REV and pore scales occur for $\phi = 0.44$ and 0.55. At higher porosity ($\phi = 0.77$ and 0.88), the RMSEs increase. However, they always remain $\lesssim 2.5 \times 10^{-3}$ Pa (Figure 10a). For a staggered array of grains (Figure 10b), the errors follow a different trend compared to the non-staggered case. In this case, the errors always remain $\lesssim 2.8 \times 10^{-3}$ Pa, and the data appear to be more scattered. A similar trend is also observed for the random array of grains (Figure 10c), where the RMSes are greater at high porosity values ($\phi = 0.77$ and 0.88).

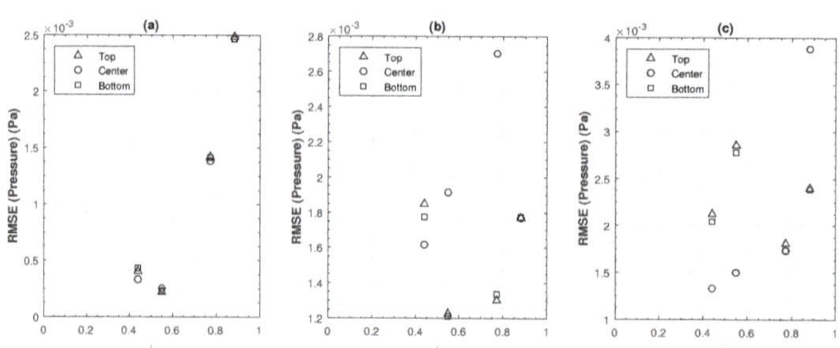

Figure 10. Root-mean-square errors (RMSEs) as functions of the porosity for the pressure losses between the REV- and pore-scale results of (**a**) Figure 7 for a non-staggered regular array of grains, (**b**) Figure 8 for a staggered regular array of grains and (**c**) Figure 9 for a randomly distributed array of grains.

The velocity profiles in Figure 5 follow similar trends to those displayed by Lai et al. [31] for the case of high porosity values $\phi \geq 0.90$. They proposed to extend the application of Ergun Equation (6) to porous media with a higher porosity and open-celled structure by simply multiplying it by a correction factor as a function of the porosity and specific surface area of the porous medium (see Equations (31) and (32)). However, when applying this correction to the case with $\phi = 0.88$, the results did not change from the $\phi = 0.88$ profiles in Figure 5. In addition, the pressure profiles in Figures 7–9 for the REV scale remained essentially the same. In this work, the REV- and pore-scale simulations were carried out using different devices. Therefore, we are not able to provide information about the computational time between both scales. In order to provide a comparison between the performance of both models, the simulations must be carried out using the same device.

6. Conclusions

In this work, the Ergun resistance force term was implemented in the momentum equation using the method of Smoothed Particle Hydrodynamics (SPH). In particular, the open-source DualSPHysics software was employed to study the flow field across a rectangular channel partially filled with a porous layer at different porosity values between 0.44 and 0.88. The flow field, which was modeled at the Representative Elementary Volume (REV) scale, was compared with similar calculations at the pore scale using three different distributions of granular solids to construct the porous medium, namely a regular non-staggered array of grains, a regular staggered array of grains and a randomly distributed array of grains. The comparison between the simulations at the REV and pore scale was made in terms of the mainstream velocity at the outlet of the rectangular channel and the head losses across both the clear and porous zones of the channel.

The main conclusions can be summarized as follows:

- The results show that the minimum differences in the flow structure and mainstream velocity profiles at the outlet of the rectangular channel between the REV- and the pore-scale simulations always occur at intermediate values of the porosity between 0.44 and 0.55.
- The distance between the REV- and the pore-scale velocity profiles at intermediate porosity as measured in terms of the root-mean-square error (RMSE) is almost independent of the geometry of the porous layer. At higher porosity, the RMSEs grow and depend more strongly on the geometry of the porous medium. However, they always remain at a very low level (i.e., $\lesssim 0.018\%$).
- The pressure losses at the REV scale reproduce those predicted at the pore scale when the grains are uniformly distributed in a non-staggered array with sufficiently good accuracy at intermediate values of the porosity (i.e., $\phi = 0.44$ and 0.55). At higher porosity, the trends of the pressure losses at the REV and pore scale are similar. In general, the distance between the profiles at the REV and pore scale increases with the porosity. However, this distance as measured in terms of the RMSE remains sufficiently small.
- The trends of the pressure losses at the REV and pore scale are affected by the geometry of the porous layer (i.e., by the grain distribution in the packed bed). The discrepancy grows for higher porosity values when the grains are randomly distributed.
- The magnitude of the pressure variations is also affected by the porosity and geometry of the porous medium. In general, the REV-scale simulations predict larger pressure losses compared to any of the pore-scale calculations.

In spite of the fact that no information is provided about the internal structure of the porous medium at the REV scale, the direct comparison with the pore-scale results in terms of the velocity field and outlet velocity profiles revealed a fairly good agreement between both scales, especially at intermediate porosity values. These results confirm that the SPH method is able to accurately describe the flow dynamics through small porous media at the REV scale. This aspect of the results becomes significantly important in larger media where the high computational load associated with such scenarios, such as, for example, in oil reservoirs, may represent an obstacle. On the other hand, the fairly good matching between the REV- and pore-scale results at relatively high porosity values makes the application of the Ergun fixed-bed equation particularly feasible in resistance calculations of porous media with higher porosity and an open-celled structure without the use of empirical corrections.

Author Contributions: The contributions for the completion of this work are listed as follows: The conceptualization was carried out jointly by all four authors. E.P. was responsible for the preparation of the models to be simulated, while L.D.-D. and C.E.A.-R. performed the numerical simulations. L.D.G.S. performed a formal analysis, wrote the manuscript and contributed with the funding acquisition. All authors have read and agreed to the published version of the manuscript.

Funding: This research was funded by the European Union's Horizon 2020 Programme under the ENERXICO Project, grant agreement No. 828947. Additionally, funding was received from the Mexican CONACYT-SENER-Hidrocarburos Programme under Grant Agreement B-S-69926.

Data Availability Statement: The experimental and numerical data supporting the conclusion of this article will be made available by the authors upon reasonable request.

Acknowledgments: One of us (C.E.A.-R.) thanks CONACYT for financial support from the Project No. 368. C.E.A.-R. is a research fellow commissioned to the University of Guanajuato (under Project No. 368). L.D.-D. thanks CONACYT for financial support through the program "Estancias Posdoctorales por México" (under grant I1200/320/2022). L.D.-D. is a postdoctoral research fellow commissioned to the Metropolitan Autonomous University, Azcapotzalco Campus. We also acknowledge financial support from the Department of Basic Sciences of the Metropolitan Autonomous University, Azcapotzalco Campus.

Conflicts of Interest: The authors declare no conflict to interest.

References

1. Hemond, H.F.; Fechner, E.J. *Chemical Fate and Transport in the Environment*; Elsevier/Academic Press: Amsterdam, The Netherlands, 2015.
2. Katopodes, N.D. *Free-Surface Flow. Environmental Fluid Mechanics*; Elsevier: Amsterdam, The Netherlands, 2019.
3. Xue, L.; Guo, X.; Chen, H. *Fluid Flow in Porous Media. Fundamentals and Applications*; World Scientific Publishing: Singapore, 2020.
4. Liu, M.; Suo, S.; Wu, J.; Gan, Y.; Hanaor, D.A.H.; Chen, C.Q. Tailoring porous media for controllable capillary flow. *J. Colloid Interface Sci.* **2019**, *539*, 379–387. [CrossRef] [PubMed]
5. Brinkman, H.C. A calculation of the viscous force exerted by a flowing fluid on a dense swarm of particles. *Appl. Sci. Res.* **1949**, *1*, 27–34. [CrossRef]
6. Liu, H.; Patil, P.R.; Narusawa, U. On Darcy-Brinkman equation: Viscous flow between two parallel plates packed with regular square arrays of cylinders. *Entropy* **2007**, *9*, 118–131. [CrossRef]
7. Rao, P.R.; Venkataraman, P. Validation of Forchheimer's law for flow through porous media with converging boundaries. *J. Hydraul. Eng.* **2000**, *126*, 63–71. [CrossRef]
8. Montillet, A. Flow through a finite packed bed of spheres: A note on the limit of applicability of the Forchheimer-type equation. *J. Fluids Eng.* **2004**, *126*, 139–143. [CrossRef]
9. Krishnan, S.; Murthy, J.Y.; Garimella, S.V. Direct simulation of transport in open-cell metal foam. *J. Heat Transf.* **2006**, *128*, 793–799. [CrossRef]
10. Gerbaux, O.; Buyens, F.; Mourzenko, V.V.; Memponteil, A.; Vabre, A.; Thovert, J.F.; Adler, P.M. Transport properties of real metallic foams. *J. Colloid Interface Sci.* **2010**, *342*, 155–165. [CrossRef]
11. Bai, M.; Chung, J.N. Analytical and numerical prediction of heat transfer and pressure drop in open-cell metal foams. *Int. J. Therm. Sci.* **2011**, *50*, 869–880. [CrossRef]
12. Mostaghimi, P.; Bijeljic, B.; Blunt, M.J. Simulation of flow and dispersion on pore-space images. *SPE J.* **2012**, *17*, 1131–1141. [CrossRef]
13. Mostaghimi, P.; Blunt, M.J.; Bijeljic, B. Computations of absolute permeability on micro-CT images. *Math. Geosci.* **2013**, *45*, 103–125. [CrossRef]
14. Piller, M.; Casagrande, D.; Schena, G.; Santini, M. Pore-scale simulation of laminar flow through porous media. *J. Phys. Conf. Ser.* **2014**, *501*, 012010. [CrossRef]
15. Mohammadmoradi, P.; Kantzas, A. Pore-scale permeability calculation using CFD and DSMC techniques. *J. Pet. Sci. Eng.* **2016**, *146*, 515–525. [CrossRef]
16. Niu, Q.; Zhang, C.; Prasad, M. A framework for pore-scale simulation of effective electrical conductivity and permittivity of porous media in the frequency range from 1 mHz to 1GHz. *J. Geophys. Res. Solid Earth* **2020**, *125*, e2020JB020515. [CrossRef]
17. Bear, J. *Dynamics of Fluids in Porous Media*; American Elsevier: New York, NY, USA, 1972.
18. Succi, S.; Foti, E.; Higuera, F. Three-dimensional floes in complex geometries with the Lattice Boltzmann method. *Europhys. Lett.* **1989**, *10*, 433–438. [CrossRef]
19. Guo, Z.; Zhao, T.S. Lattice Boltzmann model for incompressible flows through porous media. *Phys. Rev. E* **2002**, *66*, 036304. [CrossRef]
20. Seta, T.; Takegoshi, E.; Kitano, K.; Okui, K. Thermal Lattice Boltzmann model for incompressible flows through porous media. *J. Therm. Sci. Technol.* **2006**, *1*, 90–100. [CrossRef]

21. Sukop, M.C.; Huang, H.; Lin, C.L.; Deo, M.D.; Oh, K.; Miller, J.D. Distribution of multiphase fluids in porous media: Comparison between lattice Boltzmann modeling and micro-X-ray tomography. *Phys. Rev. E* **2008**, *77*, 026710. [CrossRef]
22. Zarghami, A.; di Francesco, S.; Biscarini, C. Porous substrate effects on thermal flows through a REV-scale finite volume lattice Boltzmann model. *Int. J. Mod. Phys. C* **2014**, *25*, 1350086. [CrossRef]
23. Koekemoer, A.; Luckos, A. Effect of material type and particle size distribution on pressure drop in packed beds of large particles: Extending the Ergun equation. *Fuel* **2015**, *158*, 232–238. [CrossRef]
24. He, Y.L.; Liu, Q.; Li, Q.; Tao, W.Q. Lattice Boltzmann methods for single-phase and solid-liquid phase-change heat transfer in porous media: A review. *Int. J. Heat Mass Transf.* **2019**, *129*, 160–197. [CrossRef]
25. Ashorynejad, H.R.; Zarghami, A. Magnetohydrodynamics flow and heat transfer of Cu-water nanofluid through a partially porous wavy channel. *Int. J. Heat Mass Transf.* **2018**, *119*, 247–258. [CrossRef]
26. Vijaybabu, T.R.; Dhinakaran, S. MHD natural convection around a permeable triangular cylinder inside a square enclosure filled with Al_2O_3–H_2O nanofluid: An LBM study. *Int. J. Mech. Sci.* **2019**, *153-154*, 500–516. [CrossRef]
27. Ergun, S. Fluid flow through packed columns. *Chem. Eng. Prog.* **1952**, *48*, 89–94.
28. Zhao, C.Y. Review on thermal transport in high porosity cellular metal foams with open cells. *Int. J. Heat Mass Transf.* **2012**, *55*, 3618–3632. [CrossRef]
29. du Plessis, J.P.; Woudberg, S. Pore-scale derivation of the Ergun equation to enhance its adaptability and generalization. *Chem. Eng. Prog.* **2008**, *63*, 2576–2586. [CrossRef]
30. Bazmi, M.; Hashemabadi, S.H.; Bayat, M. Modification of Ergun equation for application in trickle bed reactors randomly packed with trilobe particles using computational fluid dynamics technique. *Korean J. Chem. Eng.* **2011**, *28*, 1340–1346. [CrossRef]
31. Lai, T.; Liu, X.; Xue, S.; Xu, J. Extension of Ergun equation for the calculation of the flow resistance in porous media with higher porosity and open-celled structure. *Appl. Therm. Eng.* **2020**, *173*, 115262. [CrossRef]
32. Pang, M.; Zhang, T.; Meng, Y.; Ling, Z. Experimental study on the permeability of crushed coal medium based on the Ergun equation. *Sci. Rep.* **2021**, *11*, 23030. [CrossRef]
33. Li, Q.; Tang, R.; Wang, S.; Zou, Z. A coupled LES-LBM-IMB-DEM modeling for evaluating pressure drop of a heterogeneous alternating-layer packed bed. *Chem. Eng. J.* **2022**, *433*, 133529. [CrossRef]
34. Li, Q.; Guo, S.; Wang, S.; Zou, Z. CFD-DEM investigation on pressure drops of heterogeneous alternative-layer particle beds for low-carbon operating blast furnaces. *Metals* **2022**, *12*, 1507. [CrossRef]
35. Lucy, L.B. A numerical approach to the testing of the fission hypothesis. *Astron. J.* **1977**, *82*, 1013–1024. [CrossRef]
36. Gingold, R.A.; Monaghan, J.J. Smoothed particle hydrodynamics: Theory and application to non-spherical stars. *Mon. Not. R. Astron. Soc.* **1977**, *181*, 375–389. [CrossRef]
37. Monaghan, J.J. Smoothed particle hydrodynamics. *Rep. Prog. Phys.* **2005**, *68*, 1703–1759. [CrossRef]
38. Liu, M.B.; Liu, G.R. Smoothed particle hydrodynamics (SPH): An overview and recent developments. *Arch. Comput. Methods Eng.* **2010**, *17*, 25–76. [CrossRef]
39. Liu, G.R.; Liu, M.B. *Smoothed Particle Hydrodynamics: A Meshfree Particle Method*; World Scientific: Singapore, 2003.
40. Morris, J.P.; Zhu, Y.; Fox, P.J. Parallel simulations of pore-scale flow through porous media. *Comput. Geotech.* **1999**, *25*, 227–246. [CrossRef]
41. Zhu, Y.; Fox, P.J. Smoothed particle hydrodynamics model for diffusion through porous media. *Transp. Porous Media* **2001**, *43*, 441–471. [CrossRef]
42. Jiang, F.; Oliveira, M.S.A.; Sousa, A.C.M. Mesoscale SPH modeling of fluid flow in isotropic porous media. *Comput. Phys. Commun.* **2007**, *176*, 471–480. [CrossRef]
43. Tartakovsky, A.M.; Ward, A.L.; Meakin, P. Pore-scale simulations of drainage of heterogeneous and anisotropic porous media. *Phys. Fluids* **2007**, *19*, 103301. [CrossRef]
44. Tartakovsky, A.M.; Trask, N.; Pan, K.; Jones, B.; Pan, W.; Williams, J.R. Smoothed particle hydrodynamics and its applications for multiphase flow and reactive transport in porous media. *Comput. Geosci.* **2016**, *20*, 807–834. [CrossRef]
45. Kunz, P.; Zarikos, I.M.; Karadimitriou, N.K.; Huber, M.; Nieken, U.; Hassanizadeh, S.M. Study of multi-phase flow in porous media: Comparison of SPH simulations with micro-model experiments. *Transp. Porous Media* **2016**, *114*, 581–600. [CrossRef]
46. Kashani, M.N.; Elekaei, H.; Zivkovic, V.; Zhang, H.; Biggs, M.J. Explicit numerical simulation-based study of the hydrodynamics of micro-packed beds. *Chem. Eng. Sci.* **2016**, *145*, 71–79. [CrossRef]
47. Peng, C.; Xu, G.; Wu, W.; Yu, H.; Wang, C. Multiphase SPH modeling of free surface flow in porous media with variable porosity. *Comput. Geotech.* **2017**, *81*, 239–248. [CrossRef]
48. Lenaerts, T.; Adams, B.; Dutré, P. Porous Flow in Particle-Based Fluid Simulations. *ACM Trans. Graph.* **2008**, *27*, 1–8. [CrossRef]
49. Bui, H.H.; Nguyen, G.D. A coupled fluid-solid SPH approach to modelling flow through deformable porous media. *Int. J. Solids Struct.* **2017**, *125*, 244–264. [CrossRef]
50. Shigorina, E.; Rüdiger, F.; Tartakovsky, A.M.; Sauter, M.; Kordilla, J. Multiscale Smoothed Particle Hydrodynamics Model Development for Simulating Preferential Flow Dynamics in Fractured Porous Media. *Water Resour. Res.* **2021**, *57*, e2020WR027323. [CrossRef]
51. Bui, H.H.; Nguyen, G.D. Smoothed particle hydrodynamics (SPH) and its applications in geomechanics: From solid fracture to granular behaviour and multiphase flows in porous media. *Comput. Geotech.* **2021**, *138*, 104315. [CrossRef]

52. Nithiarasu, P.; Seetharamu, K.N.; Sundararajan, T. Natural convective heat transfer in a fluid saturated variable porosity medium. *Int. J. Heat Mass Transf.* **1997**, *40*, 3955–3967. [CrossRef]
53. Nithiarasu, P.; Seetharamu, K.N.; Sundararajan, T. Effect of porosity on natural convective heat transfer in a fluid saturated porous medium. *Int. J. Heat Fluid Flow* **1998**, *19*, 56–58. [CrossRef]
54. Becker, M.; Teschner, M. Weakly compressible SPH for free surfaces. In Proceedings of the 2007 ACM SIGGRAPH/Europhysics Symposium on Computer Animation, San Diego, CA, USA, 2–4 August 2007; Metaxas, D.; Popovic, J., Eds., pp. 32–58.
55. McCabe, W.L. *Unit Operations of Chemical Engineering. Chemical Engineering Series*, 7th ed.; McGraw-Hill: New York, NY, USA, 2005.
56. Rabbani, A.; Jamshidi, S.; Salehi, S. Determination of specific surface of rock grains by 2D imaging. *J. Geophys. Res.* **2014**, *2014*, 945387. [CrossRef]
57. Domínguez, J.M.; Fourtakas, G.; Altomare, C.; Canelas, R.B.; Tafuni, A.; García-Feal, O.; Martínez-Estévez, I.; Mokos, A.; Vacondio, R.; Crespo, A.J.C.; et al. DualSPHysics: From fluid dynamics to multiphysics problems. *Comput. Part. Mech.* **2022**, *9*, 867–895. [CrossRef]
58. Colagrossi, A.; Landrini, M. Numerical simulation of interfacial flows by smoothed particle hydrodynamics. *J. Comput. Phys.* **2003**, *191*, 448–475. [CrossRef]
59. Bonet, J.; Lok, T.S.L. Variational and momentum preservation aspects of smoothed particle hydrodynamics formulations. *Comput. Methods Appl. Mech. Eng.* **1999**, *180*, 97–115. [CrossRef]
60. Lo, E.Y.M.; Shao, S. Simulation of near-shore solitary wave mechanics by an incompressible SPH method. *Appl. Ocean. Res.* **2002**, *24*, 275–286. [CrossRef]
61. Dehnen, W.; Aly, H. Improving convergence in smoothed particle hydrodynamics simulations without pairing instability. *Mon. Not. R. Astron. Soc.* **2012**, *425*, 1068–1082. [CrossRef]
62. Alvarado-Rodríguez, C.E.; Sigalotti, L.D.G.; Klapp, J.; Fierro-Santillán, C.R.; Aragón, F.; Uribe-Ramírez, A.R. Smoothed particle hydrodynamics simulations of turbulent flow in curved pipes with different geometries: A comparison with experiments. *J. Fluids Eng.* **2021**, *143*, 091503. [CrossRef]
63. Alvarado-Rodríguez, C.E.; Klapp, J.; Sigalotti, L.D.G.; Domínguez, J.M.; de la Cruz Sánchez, E. Nonreflecting outlet boundary conditions for incompressible flows using SPH. *Comput. Fluids* **2017**, *159*, 177–188. [CrossRef]
64. Sigalotti, L.D.G.; Alvarado-Rodríguez, C.E.; Klapp, J.; Cela, J.M. Smoothed particle hydrodynamics simulations of water flow in a 90° pipe bend. *Water* **2021**, *13*, 1081. [CrossRef]
65. Morris, J.P.; Fox, P.J.; Zhu, Y. Modeling low Reynolds number incompressible flows using SPH. *J. Comput. Phyis.* **1997**, *136*, 214–226. [CrossRef]
66. Sigalotti, L.D.G.; Klapp, J.; Sira, E.; Meleán, Y.; Hasmy, A. SPH simulations of time-dependent Poiseuille flow at low Reynolds numbers. *J. Comput. Phys.* **2003**, *191*, 622–638. [CrossRef]

water

MDPI

Article

Two-Dimensional Transient Flow in a Confined Aquifer with a Cut-Off Curtain Due to Dewatering

Guangcheng Li [1], Huiming Lin [2], Min Deng [1], Lu Wang [1], Jianxiao Wang [1], Fanshui Kong [1], Yushan Zhang [1] and Qinggao Feng [2,*]

[1] Hubei Institute of Urban Geological Engineering, Wuhan 430050, China; lgctje@163.com (G.L.); melody404x@163.com (M.D.); wangluforjob@163.com (L.W.)

[2] Faculty of Engineering, China University of Geosciences, Wuhan 430074, China

* Correspondence: fengqg@cug.edu.cn

Abstract: Long, narrow, deep excavations commonly encountered in practice, such as those for subway stations, require effective groundwater management to prevent disasters in water-rich areas. To achieve this, a cut-off curtain and pumping well are typically employed during long, deep foundation pit dewatering. The unsteady groundwater flow behavior in the confined aquifer must consider the influence of the cut-off curtain during dewatering. This paper establishes a two-dimensional analytical model to describe transient groundwater flow in a confined aquifer with a cut-off curtain. Both the dewatering well pumped at a steady discharge inside the pit and the cut-off curtain are partially penetrating in the anisotropic confined aquifer. With the help of the Laplace and Fourier cosine transformations, the semi-analytical drawdown solution for the model is derived and validated against numerical solution and unsteady pumping test data. It is shown that the inserted cut-off curtain depth and the structural parameters of the pumping well significantly affect the drawdown inside the pit. Sensitivity analysis reveals that, regardless of whether the observation is made inside or outside the curtain, the drawdown is very sensitive to the change in pumping rate, aquifer thickness, storage coefficient, and horizontal hydraulic conductivity. Additionally, drawdown near the cut-off curtain outside the pit is sensitive to the vertical hydraulic conductivity of the aquifer, the width of the pit, and the interception depth of the cut-off curtain, while drawdown far from the curtain outside the pit is not sensitive to the location and length of the well screen.

Keywords: two-dimensional transient flow; semi-analytical solution; confined aquifer; cut-off curtain; sensitivity analysis

1. Introduction

The classic analytical solutions for unsteady flow in a confined aquifer due to pumping were first developed by Theis [1] for a full penetration well and by Hantush [2] for a partial penetration well. After that, the majority of well hydraulic literature has extensively examined transient flow behavior in different aquifer systems (e.g., Neuman and Witherspoon [3]; Moench [4]; Bear [5]; Perina and Lee [6]; Lin et al. [7]; Feng and Zhan [8]). The theories proposed by these studies are available in the design of a dewatering scheme for a foundation pit without a cut-off curtain in the aquifer. However, the cut-off curtain is commonly applied in deep foundation pits to enhance environmental safety and mitigate groundwater-related issues, including subsidence, sand flow, or others [9–14]. The influence of the cut-off curtain on flow behavior in aquifers during dewatering has been extensively investigated in many studies (e.g., Wu et al. [15]; Xu et al. [16]; Pujades

 https://doi.org/10.3390/w17040601

et al. [17]; Zhang et al. [18]; Wang et al. [19]; Xu et al. [20]; Li et al. [21]; Zeng et al. [22,23]; Zhang et al. [24]). These studies demonstrate that the presence of a cut-off curtain must be considered in the analysis of flow behavior in aquifers, particularly for understanding the seepage field in the vicinity of the curtain.

Many scholars have explored analytical methods for studying flow behavior in aquifer systems with cut-off curtains, providing direct and effective tools for dewatering design (e.g., Shen et al. [25]; Feng et al. [26]; Yang et al. [27]; Feng and Lin [28]). Wu et al. [29] considered the impacts of wellbore storage and well partial penetration and gave an analytical approach to analyze the characteristics of drawdown in confined aquifers with a cut-off curtain. Shen et al. [25] developed a simplified equation to compute the head difference on both sides of the curtain but failed to analyze the flow field in the entire confined aquifer. Lyu et al. [30] derived the head solutions for steady flow inside and outside the partial penetration cut-off curtain in a confined aquifer caused by a constant discharge or drawdown pumping within the pit. Feng and Lin [28] proposed an analytical approach for describing transient groundwater flow to a fully penetrating pumping well in a confined aquifer with a circular cut-off curtain. It is important to note that most of the available analytical solutions were developed on the assumption of a circular or rectangular cut-off curtain. In practical engineering, however, a long, narrow cut-off curtain is commonly constructed, especially for subway stations, deep foundation pits, and open cut tunnels in water-rich areas due to the development of urban underground spaces [31]. Under such circumstances, the flow in this long, deep excavation can be simplified into a two-dimensional flow problem in the vertical plane due to the pit's length significantly exceeding its width. However, analytical studies for dealing with two-dimensional flow in aquifers with a cut-off curtain are very limited. Tan et al. [32] proposed an analytical solution to examine steady-state groundwater flow in a confined aquifer with a cut-off curtain during the dewatering of a long-strip foundation pit. However, their study did not address the transient flow behavior in the aquifer.

To our knowledge, the analytical study on the two-dimensional transient flow due to dewatering in a confined aquifer with a partially penetrated cut-off curtain has not been investigated so far. Therefore, this study seeks to propose a semi-analytical model to describe two-dimensional transient flow in a long, deep excavation with a cut-off curtain due to dewatering in a confined aquifer. The seepage field in the whole aquifer is partitioned into two zones based on the cut-off curtain, and corresponding equations, including the governing equation and initial and boundary conditions, are established. By the use of Laplace transformation, finite Fourier cosine transformation, and discretization methods, the corresponding drawdown solution for the model is derived and subsequently validated by comparing it with both finite element method results and field pumping data. Finally, the transient drawdown response inside and outside the cut-off curtain is explored in this paper, and sensitivity analysis is performed to examine the key parameters. The proposed analytical solution provides a relatively simple, straightforward, and reliable tool to assist in the design of dewatering schemes for long foundation pit projects. It is particularly useful for the determination of some key design parameters, including the insertion depth of the cut-off curtain, as well as the number and structural configuration of dewatering wells. Additionally, the solution allows for the estimation of hydraulic parameters in the confined aquifer under the influence of the cut-off curtain.

2. Methods

2.1. Mathematical Model

Figure 1 displays a conceptual diagram of a two-dimensional flow in the vertical plane in an infinite confined aquifer with a partially penetrating well and a cut-off curtain. Note that the pumping well inside the pit is simplified as an equivalent well located at the center of the excavation. The well operates at a constant discharge and has a screen length of $(l-d)$. Additionally, the cut-off curtain partially penetrates the confined aquifer of constant thickness B, creating an open interval of length B_a. Because of the presence of the curtain, groundwater flow in the entire aquifer can be categorized into two distinct zones: Zone 1 (where $0 < x \leq x_0$) and Zone 2 (where $x_0 < x$).

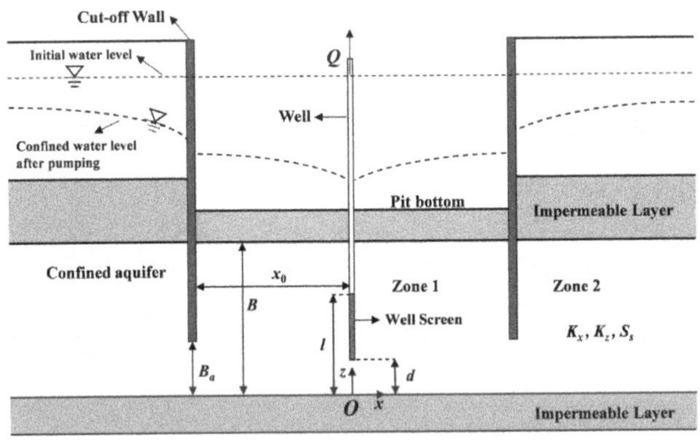

Figure 1. Sketch of the two-dimensional flow in the vertical plane in an infinite confined aquifer with a partially penetrating well and a cut-off curtain.

The assumptions are used in this study as follows (Hantush [2]; Bear [5]; Feng and Lin [28]). (1) The thickness of the cut-off curtain is negligible. (2) The pumping well has an infinitesimally small radius, and wellbore storage can be neglected. (3) The flow adheres to Darcy's law. (4) The confined aquifer is homogeneous and anisotropic.

The governing equation, along with the boundary and initial conditions for flow in Zone 1, can be written as follows:

$$K_x \frac{\partial^2 s_1(x,z,t)}{\partial x^2} + K_z \frac{\partial^2 s_1(x,z,t)}{\partial z^2} = S_s \frac{\partial s_1(x,z,t)}{\partial t}, 0 \leq x \leq x_0, 0 \leq z \leq B \tag{1}$$

$$s_1(x,z,0) = 0 \tag{2}$$

$$\lim_{x \to 0} \frac{\partial s_1(x,z,t)}{\partial x} = \begin{cases} 0, & 0 \leq z \leq d \\ \frac{-Q}{2K_x(l-d)}, & d \leq z \leq l \\ 0, & l \leq z \leq B \end{cases} \tag{3}$$

$$\frac{\partial s_1(x,0,t)}{\partial z} = \frac{\partial s_1(x,B,t)}{\partial z} = 0 \tag{4}$$

$$K_x \frac{\partial s_1(x_0,z,t)}{\partial x} = \begin{cases} q(z,t), & 0 \leq z \leq B_a \\ 0, & B_a \leq z \leq B \end{cases} \tag{5}$$

where t denotes the pumping time, Q refers to the equivalent pumping rate, K_x and K_z represent the horizontal and vertical hydraulic conductivities, respectively, and q (z, t) denotes the specific flux across the open interval. This function is continuous with respect to both z and t; s is the drawdown. Notably, Equation (3) describes the discharge condition at the well face and indicates that the flux is uniform along the well screen [2,5,26].

Similarly, the governing equation, initial and boundary conditions for flow in Zone 2 are given by:

$$K_x \frac{\partial^2 s_2(x,z,t)}{\partial x^2} + K_z \frac{\partial^2 s_2(x,z,t)}{\partial z^2} = S_s \frac{\partial s_2(x,z,t)}{\partial t}, x_0 \le x \le \infty, 0 \le z \le B \tag{6}$$

$$s_2(x,z,0) = 0 \tag{7}$$

$$s_2(\infty,z,t) = 0 \tag{8}$$

$$\frac{\partial s_2(x,0,t)}{\partial z} = \frac{\partial s_2(x,B,t)}{\partial z} = 0 \tag{9}$$

$$K_x \frac{\partial s_2(x_0,z,t)}{\partial x} = \begin{cases} q(z,t), \ 0 \le z \le B_a \\ 0, \quad B_a \le z \le B \end{cases} \tag{10}$$

where the subscripts 1 and 2 represent Zone 1 and Zone 2, respectively. Notably, Equations (5) and (10) show the flux continuity at $x = x_0$, $0 \le z \le B_a$, and additionally, the continuity of drawdown at the two zone interfaces yields:

$$s_1(x_0,z,t) = s_2(x_0,z,t), \ 0 \le z \le B_a \tag{11}$$

2.2. Solution

By first taking the Laplace transform to Equations (1)–(11) and then applying the finite Fourier cosine transform to Equations (1)–(10), the transient drawdown solutions for Zone 1 and Zone 2 can be derived using the procedure outlined in Appendix A. The final solution for Zone 1 is

$$\bar{s}_1(x,z,p) = \frac{2}{B}\sum_{n=0}^{\infty}\left\{\frac{\cosh[\eta(x_0-x)]}{\eta\sinh[\eta x_0]}\frac{Q}{2K_x(l-d)p}\delta + \frac{\cosh(\eta x)}{\eta\sinh(\eta x_0)}\frac{1}{K_x}\sum_{j=1}^{M}\bar{q}_j(p)F_j\right\}\cos\left(\frac{n\pi z}{B}\right) \tag{12}$$

where $\eta = [(K_z(n\pi/B)^2 + S_z p)/K_x]^{1/2}$, $\delta = B[\sin(n\pi l/B) - \sin(n\pi l/B)]/(n\pi)$. Notably, parameters in the Laplace domain are denoted with an overbar. p is the Laplace transform variable. Furthermore, I_0 and K_0 represent the modified Bessel functions of zeroth order for the first and second kinds, respectively, and I_1 and K_1 denote the modified Bessel functions of the first order for the first and second kinds, respectively. M signifies the quantity of discretization segments within the open interval. $\bar{q}_j(p)$ and F_j can be obtained by Equation (A22) and Equation (A23), respectively.

The solution for Zone 2 can be expressed as

$$\bar{s}_2(x,z,p) = \frac{2}{B}\sum_{n=0}^{\infty}\left\{\frac{e^{\eta(x_0-x)}}{-\eta}\frac{1}{K_x}\sum_{j=1}^{M}\bar{q}_j(p)F_j\right\}\cos\left(\frac{n\pi z}{B}\right) \tag{13}$$

Notably, the term for $n = 0$ in Equations (12) and (13) must be adjusted by multiplying it by $1/2$.

Additionally, the Stehfest [33] algorithm is employed to perform numerical Laplace inversion on the developed solutions, yielding their corresponding time–domain results.

3. Verification

3.1. Comparison with Numerical Solution

To test the validity and accuracy of the obtained solutions, we first compare the results obtained from solutions shown in Equations (12) and (13) with numerical solutions generated using the partial differential equation (PDE) module in COMSOL Multiphysics 6.0. It is important to note that the numerical scheme employed by the PDE module utilizes the finite element method rather than the finite difference method. The values used for the parameters are the following: B = 20 m, B_a = 10 m, x_0 = 20 m, l = 20 m, d = 12 m, Q = 2 m^2/d, K_x = 1 m/d, K_z = 0.5 m/d, S_s = 0.0005/m. Figure 2 shows the mesh division diagram of the numerical model with COMSOL. The model consists of 6445 free triangular elements, characterized by a maximum element size of 740 and a minimum element size of 0.0019. The simulation duration is set to 10 days, with exponentially increasing time steps resulting in a total of 1001 steps.

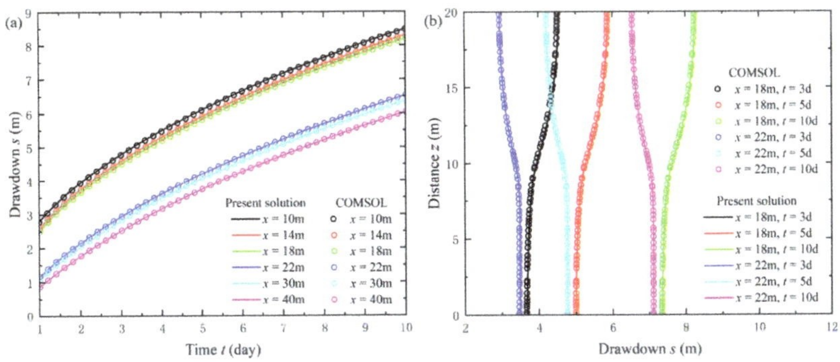

Figure 2. Comparison of drawdown obtained by the present solution (solid curve) and the numerical solution (circle symbols): (**a**) s versus t for Zone 1 (x = 10 m, 14 m and 18 m) and Zone 2 (x = 22 m, 30 m and 40 m), where z = 18 m; (**b**) z versus s at t = 3 d, 5 d and 10 d, where x = 18 m and 22 m.

Figure 2a displays the drawdown values in Zone 1 (x = 10 m, 14 m, 18 m) and Zone 2 (x = 22 m, 30 m, 40 m) at z = 18 m over the entire pumping period. Figure 2b illustrates the vertical distribution curves of drawdown for Zone 1 (x = 18 m) and Zone 2 (x = 22 m) at various pumping times (t = 3 d, 5 d, 10 d). These figures demonstrate a high degree of consistency in drawdown results between numerical simulations and analytical calculations using the developed solution, confirming the reliability and correctness of the new semi-analytical approach.

3.2. Comparison with Field Test

The developed solution can be further validated with field data of a pumping test performed in a long, deep excavation in the study of Shen et al. [27]. The flow in this pit, with dimensions of roughly 25 m in width and 152 m in length, can be simplified into a two-dimensional problem, and the detailed geological conditions and information in this site can refer to Shen et al. [27]. There are three pumping wells (Y1 to Y3) inside the pit to lower the confined water, with constant discharges of 5.42 m^3/h, 4.65 m^3/h, and 5.11 m^3/h, respectively. The equivalent total pumping discharge can be obtained as 0.1 m^2/h. The other values used in the calculations are the following [27]: B = 11 m, B_a = 0.2 m, x_0 = 12.5 m, l = 11 m, d = 4 m, K_x = 5.06 m/d, K_z = 0.51 m/d, S_s = 3 × 10^{-5}/m. Figure 3 shows the drawdowns in the observation well G2 outside the pit, calculated using the developed solution and field test data. Note that the sketch at the bottom of Figure 3 illustrates the relative position of observation well G2 in relation to the long foundation pit. It can be observed from Figure 3 that the current solution is applicable for evaluating transient drawdown, particularly for the maximum drawdown that is of interest to engineers.

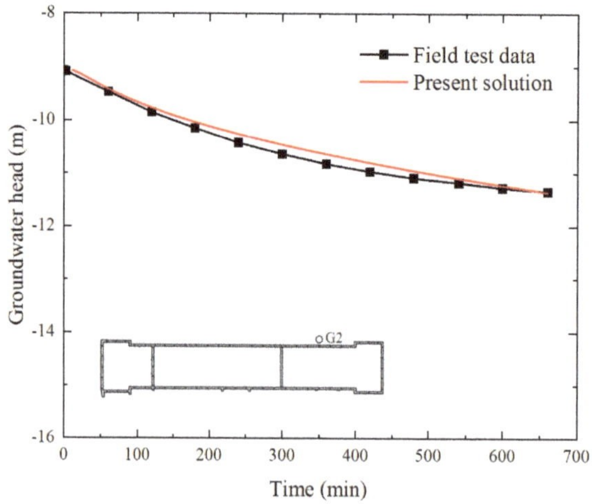

Figure 3. Groundwater head: Comparison between observed data and calculated results.

4. Results and Discussion

The drawdown characteristics inside and outside the pit are mainly investigated, and the adopted default parameters thereafter remain consistent with those employed in Section 3.1.

4.1. Effect of Curtain Insertion Depth

Figure 4 illustrates the drawdown behavior influenced by the width of the open interval B_a, ranging from 2 m to 20 m. A smaller B_a indicates a larger insertion depth of the cut-off curtain. The three subfigures in Figure 4 illustrate that variations in B_a significantly affect drawdown levels in Zone 1 and Zone 2. Specifically, Figure 4a shows that a smaller B_a leads to greater drawdown inside the curtain and reduced drawdown near the curtain outside the pit during pumping. Figure 4b demonstrates that a smaller B_a results in increased drawdown differences between the two sides of the curtain within the confined aquifer thickness. Furthermore, Figure 4c reveals that drawdown curves versus radial distance exhibit more pronounced discontinuities with greater insertion depths of the cut-off curtain. Therefore, engineers can effectively regulate groundwater levels outside the pit by adjusting the curtain insertion depth.

4.2. Effect of Well Screen Location

Figure 5 demonstrates the drawdown characteristics influenced by the well screen location, with a constant B_a = 10 m and $l - d$ = 8 m. It can be observed from Figure 5a,b that the position of the well screen within the confined aquifer significantly affects the drawdown in Zone 1, while it has a negligible impact on the drawdown in the vicinity of the roof of the confined aquifer in Zone 2. A larger value of l signifies that the well screen is positioned closer to the top of the confined aquifer. This closer proximity results in a more pronounced drawdown in Zone 1. Consequently, the obvious difference in the hydraulic head on the two sides of the cut-off curtain can be observed, as illustrated in Figure 5c. Based on the analysis above, it is recommended that the optimal location for the well screen is as close as possible to the upper boundary of the confined aquifer when designing relief wells for dewatering deep excavations inside pits. Engineers can effectively control groundwater levels within the pit by adjusting the depth at which the well screen is installed.

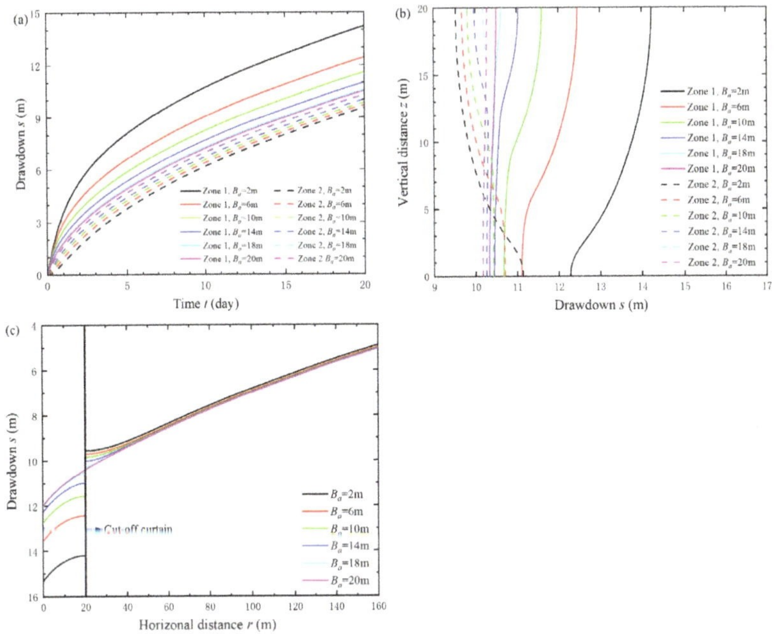

Figure 4. Curves of drawdown for various B_a (2 m, 6 m, 10 m, 14 m, 18 m, 20 m): (**a**) drawdown versus time at z = 18 m with x = 18 m for the Zone 1 and x = 22 m for Zone 2; (**b**) drawdown versus vertical distance at t = 20 d with x = 18 m for Zone 1 and x = 22 m for Zone 2; (**c**) drawdown versus horizontal distance at t = 20 d and z = 18 m.

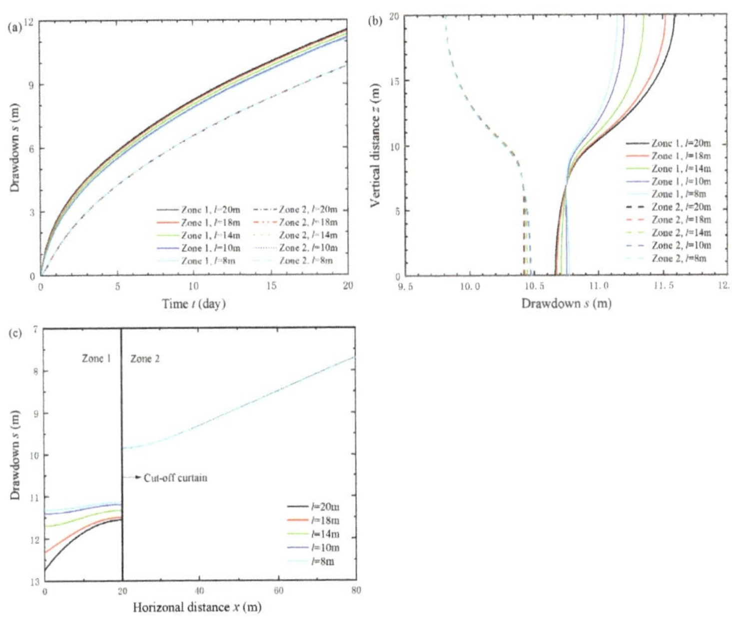

Figure 5. Curves of drawdown for various l (20 m, 18 m, 14 m, 10 m, 8 m) with B_a = 10 m: (**a**) drawdown versus time at z = 18 m with x = 18 m for Zone 1 and x = 22 m for Zone 2; (**b**) drawdown versus vertical distance at t = 20 d with x = 18 m for Zone 1 and x = 22 m for Zone 2; (**c**) drawdown versus horizontal distance at t = 20 d and z = 18 m.

4.3. Effect of Well Screen Length

Figure 6 demonstrates the drawdown responses in two zones influenced by different well screen lengths, with a fixed value of l = 20 m. Figure 6a,b shows that changes in both the length and location of the well screen have similar impacts on drawdown, particularly emphasizing a great effect on drawdown in Zone 1 compared to that in Zone 2, while the impact of the well screen length is relatively minor. A longer well screen leads to a smaller drawdown in Zone 1, indicating that it is a better choice to use a partially penetrating well in controlling the water level during dewatering design for excavation engineering. Figure 6c illustrates that variations in well screen length do not impact the horizontal distribution of drawdown outside the pit. However, there is an observable effect on the head difference on the two sides of the cut-off curtain. Specifically, increasing the length of the well screen results in a reduction in this head difference. Furthermore, for cases where d = 15 m and 10 m, implying that the well screen length $l - d \leq B_a$ = 10 m, a more pronounced drawdown within the pit is evident when $l - d \leq B_a$ compared to situations when $l - d > B_a$. This observation suggests that enhanced dewatering effectiveness within the pit can be achieved when the length of the well screen does not exceed the depth of the cut-off curtain insertion.

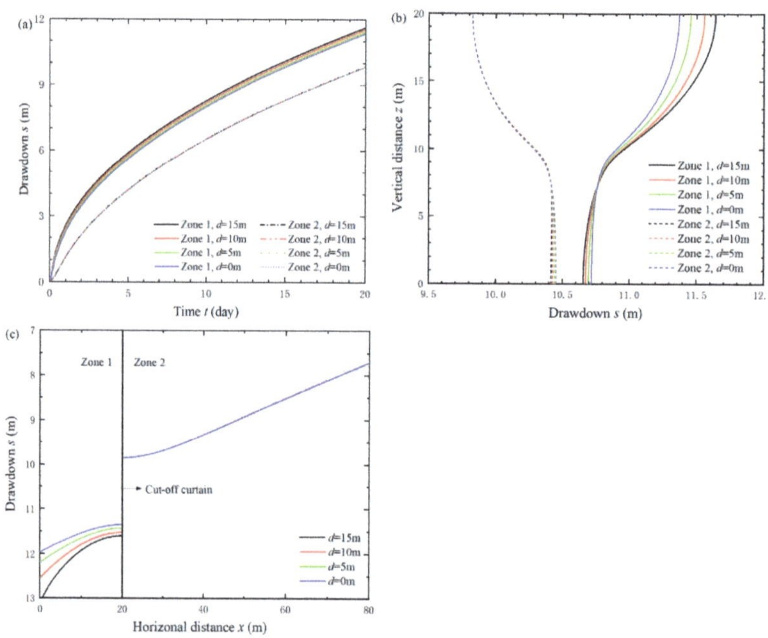

Figure 6. Curves of drawdown for various d (15 m, 10 m, 5 m, 0 m) with $B_a = l$ = 10 m: (**a**) drawdown versus time at z = 18 m with x = 18 m for Zone 1 and x = 22 m for Zone 2; (**b**) drawdown versus vertical distance at t = 20 days with x = 18 m for Zone 1 and x = 22 m for Zone 2.; (**c**) drawdown versus horizontal distance at t = 20 d and z = 18 m.

4.4. Sensitivity Analysis

Sensitivity analysis is commonly employed to evaluate the influencing degree of different hydrogeological parameters on computational outcomes [34,35]. This study adopts the sensitivity analysis method introduced by Huang and Yeh [36], which is defined as follows:

$$X_{i,j} = P_j \frac{\partial O_i}{\partial P_j} = P_j \frac{O_i(P_j + \Delta P_j) - O_i(P_j)}{\Delta P_j} \tag{14}$$

where $X_{i,j}$ refers to the normalized sensitivity coefficient (NSC) of parameter (P_j) at the j-th time step. The NSC value indicates the influence magnitude of the input parameter on the output quantity. O_i represents the dependent variable, which is the drawdown herein. ΔP_j denotes a minute increment, usually set at $10^{-2} \times P_j$. With the help of the obtained solution, one can carry out the sensitivity analysis including the following parameters: hydraulic conductivities K_x and K_z, specific storage S_s, well structure parameters l and d, open interval width B_a, and pumping discharge Q.

Figure 7 depicts the drawdown at x = 18 m, z = 18 m within the cut-off curtain to parameters B_a, K_x, K_z, S_s, B, l, d, x_0, and Q. It can be observed that the drawdown within the cut-off curtain to the pumping rate Q and well structure parameters l and d exhibits a positive correlation, and other parameters demonstrate a negative correlation, suggesting that the change of these parameters results in smaller drawdown inside the pit. Figure 7 illustrates that the drawdown within the cut-off curtain is highly responsive to Q, K_x, S_s, and B during the entire pumping process (within 20 days), and their sensitivities gradually increase with the increased pumping time, with the largest NSC for Q. Moreover, the drawdown in the pit is sensitive to K_z, x_0, and B_a and reaches a constant NSC at $t \geq 3$ d. Additionally, it is worth noting that the sensitivity of well structure parameters l, d to drawdown is primarily evident at the early stage of pumping and tends to stabilize as pumping progresses. Compared to the other parameters, the sensitivity of drawdown to l and d appears relatively weak, indicating that the impact of well structure parameters on the drawdown in the pit is not significant.

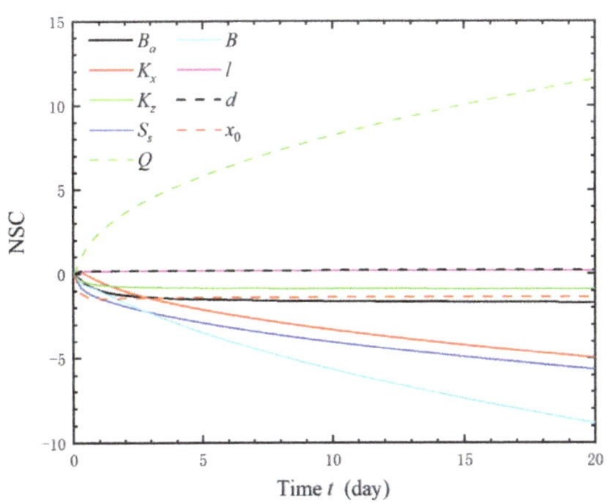

Figure 7. The NSC analysis of drawdown for the parameters B_a, K_x, K_z, S_s, B, l, d, x_0, and Q inside the cut-off curtain at x = 18 m, z = 18 m.

Figure 8 illustrates the curves of the NSC for drawdown in Zone 2 to the change in parameter B_a, K_x, K_z, S_s, B, l, d, x_0, and Q observed at z = 18 m, x = 22 m near the cut-off curtain for Figure 8a and x = 40 m away from the curtain for Figure 8b. It can be found that the pumping rate Q, vertical hydraulic conductivity K_z, and the width of the open interval B_a exhibit a positive influence on the drawdown outside the curtain, suggesting that the increase in these parameters results in a larger drawdown outside the pit. However, it is important to note that K_z has a slight influence on the drawdown, while the other parameters demonstrate a negative influence. Both Figure 8a,b demonstrate that the drawdown at any location outside the curtain is very sensitive to four parameters, Q, B, S_s, and K_x; the observation is the same as that for the drawdown in Zone 1. An important observation is that the width of the open interval B_a produces a positive influence on drawdown in Zone 2 and a negligible influence on drawdown in Zone 1. This suggests that a greater curtain insertion depth (i.e., a smaller B_a) results in a reduced drawdown outside the pit and improves the blocking effect of the curtain. Furthermore, one can observe from Figure 8a that the drawdown near the cut-off curtain outside the pit is sensitive to K_z, x_0, and B_a, while the change of these parameters has negligible influence on drawdown far away from the cut-off curtain, as displayed in Figure 8b. Finally, it can be found from Figure 8 that the change of well structure parameters l and d has no influence on drawdown outside the pit.

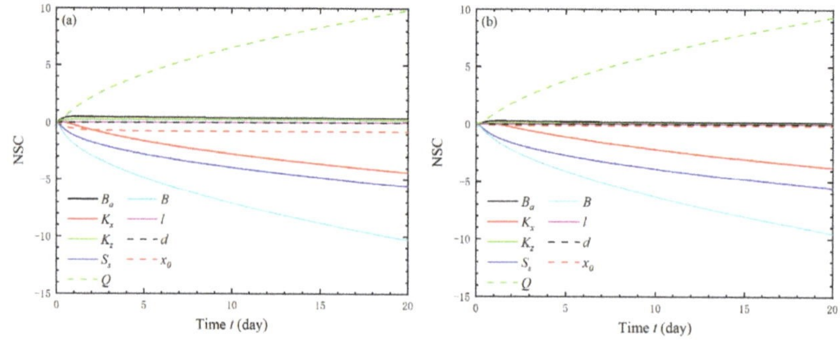

Figure 8. The NSC of drawdown outside the cut-off curtain for the parameters B_a, K_x, K_z, S_s, B, l, d, x_0, and Q (**a**) at x = 22 m, z = 18 m; (**b**) at x = 40 m, z = 18 m.

4.5. Limitations of the Present Solution

The limitations of this study are addressed herein to enhance the application of the proposed solutions. Firstly, the aquifer system is assumed to have infinite horizontal extent, a common assumption in well hydraulic literature. However, if a river or fault is near the long, narrow foundation pit, the outer boundary is typically treated under constant head or no-flow conditions. In such cases, the developed solution may not accurately represent the flow behavior in the confined aquifer. Secondly, the slope of the confined aquifer has not been considered. Thirdly, this study focuses on investigating the two-dimensional flow response in a long, narrow foundation pit; thus, the current solution cannot be applied to other geometries exhibiting significant three-dimensional flow characteristics. Fourthly, certain general limitations of the physical problem, such as uncertainties in boundary conditions and hydraulic conductivity [37,38], are not taken into account. Additionally, the non-Darcy effect of flow should be taken into account in the future. Finally, the developed solutions cannot be used to investigate groundwater flow in a heterogeneous aquifer or within an irregular spatial domain.

5. Conclusions

The purpose of this paper is to explore the transient two-dimensional flow problem in a confined aquifer with a partial penetration cut-off curtain. An analytical model is developed by accounting for the effects of aquifer anisotropy and well partial penetration. With the help of integral transformation methods, a semi-analytical solution for the model is derived. Based on the obtained solution, the unsteady-state behavior of drawdown inside and outside the cut-off curtain is then investigated. Subsequently, sensitivity analyses are carried out to determine the controlling factors, resulting in the following findings:

(1) The obtained solution is verified by making comparisons with the numerical solution using the finite element method and field pumping test data and shows its correctness and engineering applicability.

(2) The influence of the cut-off curtain on drawdown within the pit is significant compared with that outside the pit, and the drawdown difference on two sides of the curtain increases as the insertion depth of the cut-off curtain increases.
(3) To achieve better dewatering effects inside the pit, it is advisable to place the well screen near the top of the confined aquifer and maintain a screen length that is equal to or less than the insertion depth of the cut-off curtain.
(4) Regardless of whether the observation is made inside or outside the curtain, the drawdown is very sensitive to changes in parameters Q, B, S_s, and K_x, and drawdowns inside the pit and near the cut-off curtain outside the pit are sensitive to K_z, x_0, and B_a.
(5) The parameter of the open interval in the aquifer produces a positive influence on the drawdown outside the pit and has a negative effect on the drawdown inside the pit. Additionally, the well structure parameters l and d have no impact on drawdown outside the cut-off curtain.

Author Contributions: Conceptualization, G.L.; Data curation, M.D. and J.W.; Formal analysis, M.D.; Funding acquisition, G.L. and Q.F.; Investigation, L.W.; Methodology, G.L.; Project administration, G.L. and Q.F.; Software, H.L.; Supervision, Q.F.; Validation, L.W. and J.W.; Visualization, H.L.; Writing—original draft, H.L., F.K. and Y.Z.; Writing—review and editing, Q.F. All authors have read and agreed to the published version of the manuscript.

Funding: The study was supported by the National Natural Science Foundation of China (grant number 41702336) and the Science and Technology Project of the Hubei Geological Bureau (grant number KJ2022-40).

Data Availability Statement: The original contributions presented in this study are included in the article. Further inquiries can be directed to the corresponding author.

Conflicts of Interest: The authors declare no conflict of interest.

Appendix A

Applying the Laplace transformation Equations (1)–(10) yields:

$$K_x\frac{\partial^2\bar{s}_1(x,z,p)}{\partial x^2}+K_z\frac{\partial^2\bar{s}_1(x,z,p)}{\partial z^2}=S_sp\bar{s}_1(x,z,p),0\le x\le x_0,0\le z\le B \tag{A1}$$

$$\lim_{x\to 0}\frac{\partial\bar{s}_1(x,z,p)}{\partial x}=\begin{cases}0, & 0\le z\le d\\ \frac{-Q}{2K_x(l-d)}, & d\le z\le l\\ 0, & l\le z\le B\end{cases} \tag{A2}$$

$$\frac{\partial\bar{s}_1(x,0,p)}{\partial z}=\frac{\partial\bar{s}_1(x,B,p)}{\partial z}=0 \tag{A3}$$

$$K_x\frac{\partial\bar{s}_1(x_0,z,t)}{\partial x}=\begin{cases}\bar{q}(z,p), & 0\le z\le B_a\\ 0, & B_a\le z\le B\end{cases} \tag{A4}$$

$$K_x\frac{\partial^2\bar{s}_2(x,z,p)}{\partial x^2}+K_z\frac{\partial^2\bar{s}_2(x,z,p)}{\partial z^2}=S_sp\bar{s}_2(x,z,p),x_0\le x\le\infty,0\le z\le B \tag{A5}$$

$$\bar{s}_2(\infty,z,p)=0 \tag{A6}$$

$$\frac{\partial\bar{s}_2(x,0,p)}{\partial z}=\frac{\partial\bar{s}_2(x,B,p)}{\partial z}=0 \tag{A7}$$

$$K_x\frac{\partial\bar{s}_2(x_0,z,p)}{\partial x}=\begin{cases}\bar{q}(z,p), & 0\le z\le B_a\\ 0, & B_a\le z\le B\end{cases} \tag{A8}$$

$$\bar{s}_1(x_0,z,p)=\bar{s}_2(x_0,z,p),\ 0\le z\le B_a \tag{A9}$$

Then, taking the finite Fourier cosine transformation to Equations (A1)–(A8) gives

$$\frac{\partial^2 \bar{s}_{1c}(x,n,p)}{\partial x^2} - \frac{1}{K_x}\left[\left(\frac{n\pi}{B}\right)^2 K_z + S_s p\right]\bar{s}_{1c}(x,n,p) = 0,\ 0 \le x \le x_0 \tag{A10}$$

$$\lim_{x\to 0}\frac{\partial \bar{s}_{1c}(x,n,p)}{\partial x} = \frac{-Q}{2K_x(l-d)}\delta,\ d \le z \le l \tag{A11}$$

$$K_x\frac{\partial \bar{s}_{1c}(x_0,n,p)}{\partial x} = \int_0^{B_a} \bar{q}(z,p)\cos\left(\frac{n\pi z}{B}\right)dz \tag{A12}$$

$$\frac{\partial^2 \bar{s}_{2c}(x,n,p)}{\partial x^2} - \frac{1}{K_x}\left[\left(\frac{n\pi}{B}\right)^2 K_z + S_s p\right]\bar{s}_{2c}(x,n,p) = 0, x_0 \le x \le \infty \tag{A13}$$

$$\bar{s}_{2c}(\infty,n,p) = 0 \tag{A14}$$

$$K_x\frac{\partial \bar{s}_{2c}(x_0,n,p)}{\partial x} = \int_0^{B_a} \bar{q}(z,p)\cos\left(\frac{n\pi z}{B}\right)dz \tag{A15}$$

The general solution for Equations (A10) and (A13) are given by

$$\bar{s}_{1c} = A_1 e^{-\eta x} + B_1 e^{\eta x}\ \ 0 < x \le x_0 \tag{A16}$$

$$\bar{s}_{2c} = A_2 e^{-\eta x} + B_2 e^{\eta x}\ \ x_0 \le x < \infty \tag{A17}$$

Substituting Equations (A16) and (A17) into the boundary condition in Equations (A11), (A12), (A14) and (A15), A_1, B_1, A_2, and B_2 can be obtained as

$$\begin{cases} A_1 = \left[\frac{Q}{2K_x(l-d)p}\delta e^{\eta x_0} + \frac{1}{K_x}\int_0^{B_a}\bar{q}(z,p)\cos\left(\frac{n\pi z}{B}\right)dz\right]/[2\eta\sinh(\eta x_0)] \\ A_2 = \int_0^{B_a}\bar{q}(z,p)\cos\left(\frac{n\pi z}{B}\right)dz/(-K_x\eta e^{-\eta x_0}) \\ B_1 = \left[\frac{Q}{2K_x(l-d)p}\delta e^{-\eta x_0} + \frac{1}{K_x}\int_0^{B_a}\bar{q}(z,p)\cos\left(\frac{n\pi z}{B}\right)dz\right]/[2\eta\sinh(\eta x_0)] \\ B_2 = 0 \end{cases} \tag{A18}$$

where $\eta = \{[(n\pi/B)^2K_z + pS_s]/K_x\}^{1/2}$, $\delta = B[\sin(n\pi l/B) - \sin(n\pi d/B)]/(n\pi l/B)$.

Substitution of Equation (A18) into Equation (A16) and Equation (A17), respectively, and applying the inversion of the finite cos-Fourier transform leads to

$$\bar{s}_1(x,z,p) = \frac{2}{B}\sum_{n=0}^{\infty}\left\{\begin{array}{c} \frac{\cosh[\eta(x_0 - x)]}{\eta\sinh(\eta x_0)}\frac{Q}{2K_x(l-d)p}\delta \\ +\frac{\cosh(\eta x)}{\eta\sinh(\eta x_0)}\frac{1}{K_x}\int_0^{B_a}\bar{q}(z,p)\cos\left(\frac{n\pi z}{B}\right)dz \end{array}\right\}\cos\left(\frac{n\pi z}{B}\right) \tag{A19}$$

$$\bar{s}_2(x,z,p) = \frac{2}{B}\sum_{n=0}^{\infty}\left\{\frac{e^{\eta(x_0 - x)}}{-\eta}\frac{1}{K_x}\int_0^{B_a}\bar{q}(z,p)\cos\left(\frac{n\pi z}{B}\right)dz\right\}\cos\left(\frac{n\pi z}{B}\right) \tag{A20}$$

Notably, the term for $n = 0$ in Equations (A19) and (A20) is multiplied by 1/2. The unknown flux $\bar{q}(z,p)$ shown in Equations (A19) and (A20) can be resolved by Equation (A9) with the aid of the discretization approach introduced by Chang and Chen [39] and Feng and Lin [28]. Specifically, the open interval B_a is put into M segments with the length of $\Delta\xi_j$ and uniform flux $\bar{q}_j$ for segment j. Replacing the integral in Equations (A19) and (A20) with summation results in:

$$\bar{s}_1(x,z,p) = \frac{2}{B}\sum_{n=0}^{\infty}\left\{\begin{array}{c} \frac{\cosh[\eta(x_0 - x)]}{\eta\sinh[\eta x_0]}\frac{Q}{2K_x(l-d)p}\delta \\ +\frac{\cosh[\eta x]}{\eta\sinh[\eta x_0]}\frac{1}{K_x}\sum_{j=1}^{M}\bar{q}_j(p)F_j \end{array}\right\}\cos\left(\frac{n\pi z}{B}\right) \tag{A21}$$

$$\bar{s}_2(x,z,p) = \frac{2}{B}\sum_{n=0}^{\infty}\left\{\frac{e^{\eta(x_0-x)}}{-\eta}\frac{1}{K_x}\sum_{j=1}^{M}\bar{q}_j(p)F_j\right\}\cos\left(\frac{n\pi z}{B}\right) \tag{A22}$$

in which

$$F_j = \frac{2B}{n\pi}\cos\left(\frac{n\pi\bar{\xi}_j}{B}\right)\sin\left(\frac{n\pi\Delta\xi_j}{2B}\right) \tag{A23}$$

With the assumption of the average head continuity for each segment at $0 \le z \le B_a$, one can rewrite the boundary condition Equation (A9) as

$$\frac{1}{\xi_{i+}-\xi_{i-}}\int_{\xi_{i-}}^{\xi_{i+}}\bar{s}_1(x_0,z,p)dz = \frac{1}{\xi_{i+}-\xi_{i-}}\int_{\xi_{i-}}^{\xi_{i+}}\bar{s}_2(x_0,z,p)dz \tag{A24}$$

where $\xi_{j+} = \xi_j + \Delta\xi_j/2$, $\xi_{j-} = \xi_j - \Delta\xi_j/2$ and $\Delta\xi_j = \xi_{j+} - \xi_{j-}$.

Substituting Equations (A21) and (A22) into Equation (A24) leads to

$$\sum_{i=1}^{M}\bar{q}_j(p)X_{ij} = Y_i \tag{A25}$$

where

$$X_{ij} = \frac{1}{K_x}\left[\frac{2}{B}\sum_{n=0}^{\infty}\frac{\cosh(\eta x_0)}{\eta\sinh(\eta x_0)}F_jF_i + \frac{2}{B}\sum_{n=0}^{\infty}\frac{1}{\eta}F_iF_j\right] \tag{A26}$$

$$Y_i = -\frac{2}{B}\sum_{n=0}^{\infty}\left[\frac{1}{\eta\sinh(\eta x_0)}\frac{Q}{2K_x(l-d)p}\delta\right]F_i \tag{A27}$$

$$F_i = \frac{2B}{n\pi}\cos\left(\frac{n\pi\bar{\xi}_i}{B}\right)\sin\left(\frac{n\pi\Delta\xi_i}{2B}\right) \tag{A28}$$

References

1. Theis, C.V. The relation between the lowering of the piezometric surface and the rate and duration of discharge of a well using ground-water storage. *Trans. Am. Geophys. Union* **1935**, *16*, 519–524. [CrossRef]
2. Hantush, M.S. Hydraulics of wells. *Adv. Hydrosci.* **1964**, *1*, 281–432.
3. Neuman, S.P.; Witherspoon, P. Theory of flow in a confined two aquifer system. *Water Resour. Res.* **1969**, *5*, 803–816. [CrossRef]
4. Moench, A.F. Transient flow to a large-diameter well in an aquifer with storative semiconfining layers. *Water Resour. Res.* **1985**, *21*, 1121–1131. [CrossRef]
5. Bear, J. *Hydraulics of Groundwater*; McGraw-Hill: Dover, NY, USA, 2007.
6. Perina, T.; Lee, T.C. General well function for pumping from a confined, leaky, or unconfined aquifer. *J. Hydrol.* **2006**, *317*, 239–260. [CrossRef]
7. Lin, Y.C.; Yang, S.Y.; Fen, C.S.; Yeh, H.D. A general analytical model for pumping tests in radial finite two-zone confined aquifers with Robin-type outer boundary. *J. Hydrol.* **2016**, *540*, 1162–1175. [CrossRef]
8. Feng, Q.G.; Zhan, H.B. Two-region flow caused by pumping at a partial penetration well in a leaky confined aquifer. *Int. J. Numer. Anal. Methods Geomech.* **2023**, *47*, 717–735. [CrossRef]
9. Pujades, E.; Carrera, J.; Jurado, A.; Vázquez-Suñé, E.; Vilarrasa, V.; Mascuñano-Salvador, E. Hydraulic characterization of diaphragm walls for cut and cover tunnelling. *Eng. Geol.* **2012**, *125*, 1–10. [CrossRef]
10. Shen, S.L.; Xu, Y.S. Numerical evaluation of land subsidence induced by groundwater pumping. *Can. Geotech. J.* **2011**, *48*, 1378–1392. [CrossRef]
11. Tan, Y.; Lu, Y.; Wang, D.L. Deep excavation of the gate of the orient in Suzhou stiff clay: Composite earth-retaining systems and dewatering plans. *J. Geotech. Geoenviron.* **2018**, *144*, 05017009. [CrossRef]
12. Wang, J.X.; Liu, X.T.; Liu, S.L.; Zhu, Y.F.; Pan, W.Q.; Zhou, J. Physical model test of transparent soil on coupling effect of cut-off wall and pumping wells during foundation pit dewatering. *Acta Geotech.* **2019**, *14*, 141–162. [CrossRef]
13. Zeng, C.F.; Wang, S.; Xue, X.L.; Zheng, G.; Mei, G.X. Evolution of deep ground settlement subject to groundwater drawdown during dewatering in a multi-layered aquifer-aquitard system: Insights from numerical modelling. *J. Hydrol.* **2021**, *603*, 127078. [CrossRef]

14. Zheng, G.; Cao, J.R.; Cheng, X.S.; Ha, D.; Wang, F.J. Experimental study on the artificial recharge of semiconfined aquifers involved in deep excavation engineering. *J. Hydrol.* **2018**, *557*, 868–877. [CrossRef]
15. Wu, Y.X.; Shen, S.L.; Yuan, D.J. Characteristics of dewatering induced drawdown curve under blocking effect of retaining wall in aquifer. *J. Hydrol.* **2016**, *539*, 554–566. [CrossRef]
16. Xu, Y.S.; Shen, S.L.; Ma, L.; Sun, W.J.; Yin, Z.Y. Evaluation of the blocking effect of retaining walls on groundwater seepage in aquifers with different insertion depths. *Eng. Geol.* **2014**, *183*, 254–264. [CrossRef]
17. Pujades, E.; Vazquez-Sune, E.; Carrera, J.; Jurado, A. Dewatering of a deep excavation undertaken in a layered soil. *Eng. Geol.* **2014**, *178*, 15–27. [CrossRef]
18. Zhang, Y.Q.; Wang, J.H.; Chen, J.J.; Li, M.G. Numerical study on the responses of groundwater and strata to pumping and recharge in a deep confined aquifer. *J. Hydrol.* **2017**, *548*, 342–352. [CrossRef]
19. Wang, X.W.; Yang, T.L.; Xu, Y.S.; Shen, S.L. Evaluation of optimized depth of waterproof curtain to mitigate negative impacts during dewatering. *J. Hydrol.* **2019**, *577*, 123969. [CrossRef]
20. Xu, Y.S.; Yan, X.X.; Shen, S.L.; Zhou, A.N. Experimental investigation on the blocking of groundwater seepage from a waterproof curtain during pumped dewatering in an excavation. *Hydrogeol. J.* **2019**, *27*, 2659–2672. [CrossRef]
21. Li, M.G.; Chen, J.J.; Xia, X.H.; Zhang, Y.Q.; Wang, D.F. Statistical and hydro-mechanical coupling analyses on groundwater drawdown and soil deformation caused by dewatering in a multi-aquifer-aquitard system. *J. Hydrol.* **2020**, *589*, 125365. [CrossRef]
22. Zeng, C.F.; Liao, H.; Xue, X.L.; Long, S.C.; Luo, G.J.; Diao, Y.; Li, M.G. Responses of groundwater and soil to dewatering considering the barrier effect of adjacent metro station on multi-aquifers. *J. Hydrol.* **2022**, *612*, 128117. [CrossRef]
23. Zeng, C.F.; Chen, H.B.; Liao, H.; Xue, X.L.; Chen, Q.N.; Diao, Y. Behaviours of Groundwater and Strata during Dewatering of Large-Scale Excavations with a Nearby Underground Barrier. *J. Hydrol.* **2023**, *20*, 129400. [CrossRef]
24. Zhang, X.H.; Wang, X.W.; Xu, Y.S. Influence of filter tube of pumping well on groundwater drawdown during deep foundation pit dewatering. *Water* **2021**, *13*, 3297. [CrossRef]
25. Shen, S.L.; Wu, Y.X.; Misra, A. Calculation of head difference at two sides of a cutoff barrier during excavation dewatering. *Comput. Geotech.* **2017**, *91*, 192–202. [CrossRef]
26. Feng, Q.G.; Cai, B.H.; Feng, X.L.; Yuan, X. Analytical solutions of transient flow model for a partially penetrating well in a finite leaky confined aquifer system. *Rock Soil Mech.* **2021**, *42*, 168–176.
27. Yang, K.F.; Xu, C.J.; Wang, P. Analytical analysis of the groundwater drawdown difference induced by foundation pit dewatering with a suspended cut-off curtain. *Appl. Sci.* **2022**, *12*, 10301. [CrossRef]
28. Feng, Q.G.; Lin, H.M. Transient flow induced by pumping in an anisotropic confined aquifer with a cut-off curtain during excavation dewatering. *Comput. Geotech.* **2024**, *168*, 106116. [CrossRef]
29. Wu, Y.X.; Shen, J.S.; Chen, W.C.; Hino, T. Semi-analytical solution to pumping test data with barrier, wellbore storage, and partial penetration effects. *Eng. Geol.* **2017**, *226*, 44–51. [CrossRef]
30. Lyu, H.M.; Shen, S.L.; Wu, Y.X.; Zhou, A.N. Calculation of groundwater head distribution with a close barrier during excavation dewatering in confined aquifer. *Geosci. Front.* **2021**, *12*, 791–803. [CrossRef]
31. Sun, X.Y.; Jiang, Z.X.; Tong, L.Y.; Sun, J.W.; Cui, J.; Zhou, X.; Liu, S.Y. Optimization of confined aquifer dewatering for long-deep excavation using simulation-optimization method. *Undergr. Space* **2024**, *17*, 246–266. [CrossRef]
32. Tan, W.J.; Kang, H.B.; Xu, J.; Wang, X.D. Analysis of steady groundwater flow in confined aquifer due to long-strip pit dewatering with suspended cut-off curtain. *Sustainability* **2023**, *15*, 15699. [CrossRef]
33. Stehfest, H. Numerical inversion of Laplace transforms. *Commun. ACM* **1970**, *13*, 47–49. [CrossRef]
34. Kabala, Z.J. Sensitivity analysis of a pumping test on a well with wellbore storage and skin. *Adv. Water Resour.* **2001**, *24*, 483–504. [CrossRef]
35. Feng, Q.G.; Luo, Y.; Zhan, H.B. Three-dimensional response to a partially penetration well pumping in a general anisotropic three-layer aquifer system. *J. Hydrol.* **2020**, *585*, 124850. [CrossRef]
36. Huang, Y.C.; Yeh, H.D. The use of sensitivity analysis in on-line aquifer parameter estimation. *J. Hydrol.* **2007**, *335*, 406–418. [CrossRef]
37. Samarinas, N.; Tzimopoulos, C.; Evangelides, C. Fuzzy Numerical Solution to the Unconfined Aquifer Problem under the Boussinesq Equation. *Water Supply* **2021**, *21*, 3210–3224. [CrossRef]
38. Tzimopoulos, C.; Papadopoulos, K.; Samarinas, N.; Papadopoulos, B.; Evangelides, C. Fuzzy Finite Elements Solution Describing Recession Flow in Unconfined Aquifers. *Hydrology* **2024**, *11*, 47. [CrossRef]
39. Chang, C.C.; Chen, C.S. A flowing partially penetrating well in a finite-thickness aquifer: A mixed-type initial boundary value problem. *J. Hydrol.* **2003**, *271*, 101–118. [CrossRef]

MDPI AG
Grosspeteranlage 5
4052 Basel
Switzerland
Tel.: +41 61 683 77 34

Water Editorial Office
E-mail: water@mdpi.com
www.mdpi.com/journal/water